Introduction to the economics of water resources

Introduction to the economics of water resources

AN INTERNATIONAL PERSPECTIVE

Stephen Merrett

UCL PRESS

First published in 1997 by UCL Press

UCL Press Limited
1 Gunpowder Square
London EC4A 3DE

and

1900 Frost Road, Suite 101
Bristol
Pennsylvania 19007-1598
USA

The name of University College London (UCL) is a registered
trade mark used by UCL Press with the consent of the owner.

British Library Cataloguing in Publication Data
A CIP catalogue record for this book is available from the British Library.

Library of Congress Cataloging-in-Publication Data are available.

ISBNs: 1-85728-636-7 HB
       1-85728-637-5 PB

Printed and bound in Great Britain.
By T. J. International Ltd, Padstow, Cornwall.

CONTENTS

This work is dedicated with love to my wife Alicia,
to my daughters Selena and Juley,
and to the memory of my mother Ada Elizabeth Sinclair

ACKNOWLEDGEMENTS

I wish to acknowledge the invaluable advice and commentary received on draft chapters of the book from Tony Allan, Brian Arkell, Henning Björnlund, Alison Brook, Judy Day, John Farrant, Danka Heliová, Peter Hlavatý, Geoff Hodgson, Mike Horner, Jennifer McKay, Milan Matuška, Max Neutze, John Pigram, Kirsty Powell, Manuel Schiffler, Richard Streeter, Erik Swyngedouw, Nick Walton and Jim Wescoat.

ABBREVIATIONS AND ACRONYMS

af	acre-feet
APC	average prime cost
ATC	average total cost
ATCmin	minimum average total cost
ATP	Aid and Trade Provision
BATNEEC	best available technique not entailing excessive cost
BOD	biological oxygen demand
BOD5	BOD measured over five days at 20°C
CEC	Commission of the European Communities
cl	centilitre
cm	centimetre
CSERGE	Centre for Social and Economic Research on the Global Environment
dl	decilitre
dm	decimetre
DTI	Department of Trade and Industry
EAEPE	European Association for Evolutionary Political Economy
EBRD	European Bank for Reconstruction and Development
ECGD	Export Credits Guarantee Department
EIA	environmental impact assessment
EPDRB	Environmental Programme for the Danube River Basin
EU	European Union
F	Farad
g	gram
GDP	gross domestic product
GEF	global environmental facility
GMIA	Goulburn–Murray Irrigation Area
h	hecto
ha	hectare
HOCCPA	House of Commons Committee of Public Accounts
HOCFAC	House of Commons Foreign Affairs Committee
Hz	hertz
IADB	Inter-American Development Bank
IRR	internal rate of return
k	kilo
kg	kilogram
km	kilometre

kV	kilovolt
l	litre
l per day	litres per day
m	metre
mg per l	milligrams per litre
ml	millilitre
mm	millimetre
mm per year	millimetres per year
M	mega
Ml per day	megalitres per day
MW	megawatt
m^3	cubic metre
m^3 per day	cubic metres per day
m^3 per s	cubic metres per second
m^3 per year	cubic metres per year
NAO	National Audit Office
NGO	non-government organization
NPV	net present value
NRA	National Rivers Authority
n/a	not applicable
ODA	Overseas Development Administration
Ofwat	Office of Water Services
plc	public limited company
p per m^3	pence per cubic metre
pppd	per person per day
PRONAP	Programa Nacional de Agua Potable y Alcantarillado
RSPB	Royal Society for the Protection of Birds
SCBA	social cost–benefit analysis
SCEA	social cost-effectiveness analysis
SIMOP	modelo de simulación de obras públicas
SMEC	Snowy Mountains Engineering Corporation
STRD	social time rate of discount
TOC	total organic carbon
UK	United Kingdom
UNCED	United Nations Conference on Environment and Development
US	United States (adjective)
USA	United States of America
USSR	Union of Soviet Socialist Republics
V	volt
W	watt

UNITS OF MEASUREMENT

deca   10
hecto 10^2
kilo 10^3
mega 10^6
deci 10^{-1}
centi 10^{-2}
milli 10^{-3}
micro 10^{-6}

acre   An antique English measure of area. 1 acre equals 0.405 hectares.
acre-foot   An antique English measure of volume. 1 acre-foot equals $1182\,m^3$.
gram (also gramme)   A metric unit of mass.
litre   A metric unit of volume, equal to one cubic decimetre.
metre*   A metric unit of length.
metre–kilogram–second   Denoting a system of units of measurement using the metre, kilogram and second as the basic units of length, mass and time.
SI (Système International)   The international system of units of measurement.
tonne   1000 kilograms.

* Note that one cubic metre is equal to 1000 litres.

CHAPTER ONE

Introduction

Sunlight, air, the soil and water – these are the fundamental requirements of all life on Earth. In the specific case of water, the human body cannot survive without it, it plays a vital part in sanitation for our rural and urban communities, it is necessary to all forms of agriculture, and is demanded for the majority of industrial processes. So, water is a key natural resource for human society.

Unlike sunlight and air, we know that rivers, lakes, estuaries and coastal waters can all be appropriated into the ownership of public or private bodies. For this reason alone, water is not merely a natural resource but also an economic resource. This is true even though in many countries no price is set on water's use, no sum of money charged per unit consumed. This is the case whenever users enjoy unrestricted physical access to fresh water; and also where water is supplied to users by public or private water companies but its volume is not measured. In the latter case, water charges are widely collected by means of a fixed charge levied on water consumers.

Priced or not, we should still recognize that for most societies the collection and distribution of fresh water requires human labour and, often, civil engineering infrastructures, as was the case in ancient Egypt, Imperial China and the Inca civilization. So, for this reason too, water is and always has been an economic resource. In more recent times, this understanding has become much more widely accepted because of the economic and financial costs imposed by laws to protect water quality; because of water scarcity and the associated competition between users; and as a result of the global shift to the privatization of public sector infrastructures since the end of the 1970s.

If water is so fundamental a biological and social requirement, and if it is now widely recognized to be an economic good, then there is a need for hydroeconomics – the economics of water resources – and there is a potentially varied audience with an interest in that subject. This book is intended to meet the needs of both students and professionals in the fields of economics, engineering, environmental science, environmental studies, geography and hydrology. As an introduction to the subject, prior knowledge is assumed of neither economics nor hydrology. To facilitate the learning process, the analytical content of each of the next seven chapters is always followed by one or more case studies, drawing their material from countries as diverse as the UK, Slovakia, Latvia, the USA, Peru, Jordan, Malaysia and Australia.

If, then, a political economy of water resources exists, what may be said to be its subject areas, that is, what human activities does it address? Since no grand academy determines the content of the subdisciplines of economics, the answer to this question embodies an element of personal judgement. I shall define the subject areas of hydroeconomics as:

- nature conservation for rivers, lakes, wetlands, estuaries and coastal waters
- land drainage
- flood control and coastal defence
- dam projects
- the supply of fresh water
- the use of water by households, agriculture, industry and other sectors
- the treatment of wastewater and its disposal.

Note that this definition does not include fishing, navigation or water-based recreation. Although the hydroeconomist will require some empathy with these areas, they can best be regarded as components of the economics of fishing, transport and leisure respectively. At the same time, hydroeconomists can expect their passion for nature conservation to be shared with environmental economists, the analysis of dams to be strengthened by familiarity with the economics of energy, and the study of the demand for water by farmers to draw fruitfully on the economics of agriculture. Human culture is never neatly separable into discrete dimensions, nor is political economy. To restate the position, the primary orientation of the economics of water resources is to water that is abstracted, stored and distributed by human labour, to the use of that water, and to the disposal of wastewater.

With the subject areas of hydroeconomics specified, it is possible to formulate its substantive concerns. Once again, an element of personal judgement is necessary. The substantive concerns of hydroeconomics, which can be expressed as the guiding criteria it requires for the development of analysis and policy, are as follows:

- to supply water of sufficient quantity and appropriate quality to users in households, agriculture, industry and other sectors
- to ensure the use of fresh water is affordable to low-income households
- to ensure the husbandry of water in its supply and use
- to purify water from domestic, agricultural and industrial effluents
- to prevent the abuse of monopoly power in the supply of fresh water and the collection of wastewater
- to protect rural and urban communities against floods and to drain the land of stormwater
- to protect water's hydrocyclical capacity to renew its ground- and surface-water flows
- to conserve natural species and habitats in all their fresh- and coastal-water environments

- to reduce and eliminate water-driven international conflict
- to ensure that, when government expenditure takes place for these purposes, it is spent wisely.

The economic paradigm that provides the analytical foundation of this book deserves mention. This is evolutionary political economy, rather than Marxist economics or neoclassical economics – the two dominant economic paradigms of this century.[1] An important expositional text is Hodgson et al. (1994). It is used hereafter as a reference point throughout my own book, which attempts, for the first time, to apply evolutionary political economy to the supply and use of water resources and which, at various junctures in the text, also develops specific criticisms of neoclassical orthodoxy. Hodgson et al. (1994) is a good starting point for those who may wish to explore evolutionary political economy further.

The book's analytical structure is as follows. Chapter 2 gives an overview, from a physical infrastructural point of view, of the supply of fresh- and wastewater services. Chapter 3 follows with an account of the costs of supplying these services. Chapter 4 complements the two initial supply-side perspectives with a review of the economics of effective demand and of the price of water. In Chapter 5, an exposition is given of the techniques of social cost–benefit analysis with respect to water projects, and Chapter 6 matches this with an outline of financial accounting for water enterprises. Chapters 7 and 8 are concerned respectively with water's role in achieving a sustainable society and the valuation of the environmental costs and benefits of water projects. The final chapter gathers together the principal conclusions of the whole work. From this brief listing of chapters it can be seen that the central analytical themes developed in addressing the substantive concerns of hydroeconomics are supply, effective demand, project analysis, enterprise finance and the sustainable management of water resources.

The main body of the text is followed by a glossary, the references and a useful index. From Chapter 2 onwards where a term will be placed in the glossary, it is formatted in bold the first time it occurs in the text. Please note, too, that in the presentation of numbers, English practice is used. For example, 1243 is one thousand, two hundred and forty-three, whereas 1.243 is a decimal expression.

---

1. In Europe, the professional association committed to the development of the paradigm is the European Association for Evolutionary Political Economy (EAEPE).

CHAPTER TWO
Supply: the engineer's perspective

2.1 The hydrological cycle

Evolutionary political economy focuses on the substance of the fundamental processes involved in providing necessary goods and services for humankind. So, rather than make a beginning with the economic theory of supply, here we start with an overview of the natural world's **hydrological cycle**, moving on to the engineering activities of what I shall call the **hydrosocial cycle** in the supply of fresh- and wastewater services.

Water held in the oceans and other water bodies, as well as in the land masses and their vegetation, continuously evaporates into the global atmosphere, whence it returns to the Earth through precipitation, primarily as rainfall. On land it gathers as **surface water** and infiltrates as groundwater. Surface water is the flow of streams and rivers in spatially distinct **catchment** areas, as well as their associated freshwater pools, wetlands, lakes, inland seas and river deltas. Groundwater is found below the land's surface, and includes that contained in **aquifers**. Such aquifers help to sustain surface-water flows. Of all the world's water, 97.4 per cent is in the oceans and 2.6 per cent is land-based, with an additional tiny fraction held in the atmosphere. Of all the land's water, 76.4 per cent is in ice caps and glaciers, 22.8 per cent occurs as groundwater, and 0.6 per cent is surface water, half of it in saline seas (Ward & Robinson 1990: table 1.1).

Water can be measured in stock terms, that is, the volume present in a particular place at a point in time. Examples are the size of an aquifer, or of a lake, estimated in cubic metres. Water quantities are also calculated in flow terms, the volume reaching or passing a defined point or area in any time period. Examples are a spring's flow in litres per second, or a river's flow in cubic metres per day, or rainfall in a region in millimetres depth per year. The term "capacity" in water planning is sometimes used to refer to a stock of water, as with "the capacity of a **reservoir**", and sometimes to a flow, as with the annual capacity of a **water treatment plant**.

2.2 The hydrosocial cycle

The engineering activities of the hydrosocial cycle in the supply of fresh- and wastewater services are illustrated in Figure 2.1. Seven boxes there are of special importance: **abstraction**, **storage**, freshwater treatment, freshwater **distribution**, wastewater collection, wastewater treatment, and wastewater disposal. Each of these processes will be considered in turn, as well as the demand-side activity of **consumption**, which lies between freshwater distribution and wastewater collection.

A starting point is the natural flow in a catchment which is available for the conversion from natural resource to social product. This flow varies during the course of the year because of the changing seasons. Moreover, there is year-on-year variation: an abundance of rain in one 12-month period may be succeeded by long, dry spells in the next. Of course, these seasonal and annual supply variations differ markedly between regions and countries.

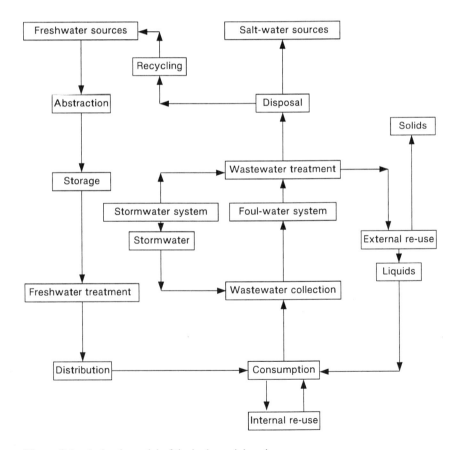

Figure 2.1 A simple model of the hydrosocial cycle.

The theoretically available resource is referred to as effective rainfall, that is, total rainfall in an area minus that lost both through **transpiration** by foliage and through surface **evaporation**. In a supply forecasting context, widely used measures are the effective rainfall during average rainfall years, as well as effective drought rainfall, that available in a drought of a severity which could be expected to recur, for example, once in every 50 years.

Abstraction takes place from groundwater, from freshwater surface sources, from saline inland seas, from tidal sources and from the open sea. In these last three cases (of saline water) our interest is where **brackish** estuarine water is used for specific purposes such as cooling in electricity generation, or where sea water is treated in **desalination** plants.

The abstraction of fresh water may be directly by the user, such as in a riverside village, on a farm, in an industrial firm or for a hydroelectric power station; or it may be by a water company, which supplies the water as a service. Water companies can be in public or private ownership. In the European Union (EU), for example, direct users as well as water companies must have an abstraction licence, which defines the maximum flow which it is permissible to take.

Abstraction requires source works. In the case of groundwater, **boreholes** are driven down into the aquifer, and pumping equipment is installed to get the water to the starting point of the hydrosocial network. Accessing surface water is easier but still requires source infrastructures.

There is more than one definition of abstraction capacity. Licensed abstractions have already been referred to. **Installed capacity** is the flow of water that it is theoretically possible to collect with the existing technical infrastructures. **Usable capacity** is lower than this whenever what is actually possible to abstract falls short of the theoretical limit. This shortfall may result from equipment malfunctions attributable to age or inadequate maintenance. This distinction between installed and usable capacity is often used, for example, in water resource planning in eastern Europe. Also, the **safe yield** of installed capacity is often used; that is, the flow of water available, for example, during a 1 in 50 year drought. In addition, it is always necessary to have spare stand-by capacity to safeguard supplies during breakdowns and maintenance periods.

Actual abstraction is the flow collected from a water source in practice. This may well fall short of authorized abstraction, because the higher authorized volume is not required by consumers. The reassignment of authorizations between abstractors within a single catchment area is one way to increase the actual total abstracted from a given river system. However, where saline abstraction licences are not fully utilized, they are of little value elsewhere.

In some cases, actual abstractions may fall within their licensed maximum, yet exceed the effective rainfall in a specific year. Over-abstraction can lead to low river flow as the **water table** falls. Unnatural flow regimes influence stream ecology and, in these circumstances, create a conflict between the needs of abstractors and those of the water environment.

2.3 Storage

The preceding section looked at the first of a fresh- and wastewater supply system's seven fundamental processes, that of abstraction. Now comes the turn of the second, storage.

The inevitable variation in effective rainfall within the year and between years makes the natural supply of water inconstant. In dry spells and in droughts, this has the effect of reducing actual abstractions precisely when farmers and households may wish to consume more water. Water demand also fluctuates with the level of economic activity. Without some buffering device, the effect of this variance in supply and demand would be to create local water shortages, sometimes with disastrous consequences. Water storage moderates these difficulties by providing a stock of water for times when effective rainfall is low in relation to demand. Storage takes the form of reservoirs and water tanks.

It should be stressed that Figure 2.1 is a simplification of what are often extremely complex networks. Water for urban uses is often stored after treatment as well as prior to it. These storage points are at relatively high elevation, the water towers we so often see on the skyline, raising pressure for the consumer at the point of delivery. They are also important in coping economically with **diurnal** fluctuations in use. Bulk **mains** carry water from the main storage and treatment plants to these local balancing storages at a steady rate throughout the day and night, without having to be large enough themselves to cope with peak demand during the day (Neutze 1997).

Reservoirs are essential elements in the integration of catchment supply networks. Rees with Williams (1993: 75) have pointed out how they can be managed to increase output from existing infrastructures. In the case of the UK's National Rivers Authority (NRA) Yorkshire region, it has been argued that a more sophisticated control system for the management of upland reservoirs could yield an extra 20 million litres per day. Moreover, the conjunctive use of river, underground and upland reservoir sources would bring even higher gains, for example by abstracting more intensively from rivers in the winter, when flows are strong, so reserving storage and groundwater supplies for the summer months. Water storage often has an associated function as a store of potential energy, and this may lead to conflict over release timings between electricity generating companies and other water users.

Now we consider some natural forms of freshwater storage. Glaciers and icecaps are good temporary stores of excess winter precipitation, slowly releasing their supplies through the spring and early summer, just when rivers would be running low and crops need water. Underground aquifers are major natural storage places. For example, $10^{12} m^3$ is stored in UK chalk deposits. They allow effective storage of excess winter **recharge** for access during dry summers. Artificial recharge permits the storage of excessive winter river flow into aquifers for later use, as in the case of the London recharge scheme.

2.4 Freshwater treatment

The third major building block of the hydrosocial cycle is freshwater treatment to remove both undesirable natural substances and man-made contaminants. This introduces the concept of water quality to place alongside that of water quantity. A special case of treatment, not strictly of a contaminant, is the desalination of sea water to generate freshwater supplies. Desalination is energy-intensive and is found largely in the countries of the Middle East and the US Sunshine Belt. Treatment can also include the addition of soluble elements such as fluoride, to enhance water in the interests of public health.

Pollution of the natural world, including its freshwater resources, is the result of human society's activities of production and consumption. The use of water results in its degradation. Often, this can be gradual and, in such cases, it becomes contaminated only through repeated use. Among hydrochemists, the term "polluted water" implies a high degree of contamination. So:

degradation → contamination → pollution

is a way of indicating increasingly poor levels of water quality.

Water quality is assessed by its physical, biological and chemical characteristics, and contamination radically changes any or all of these. It may originate from **point** or from **ambient** sources and can affect both groundwater and surface water. The main forms of contamination in Europe are **sewage** discharge **pathogens**, both bacterial and viral, nitrates from fertilizer use, heavy metals from soil and urban runoff, mineral oil discharges from illegal dumping, chlorinated solvent discharges from poorly managed waste-disposal sites, acid rain, and a cocktail of poisons from working (or abandoned) industrial and mineral sites. Water pollutants are usually measured in micrograms per litre.

Freshwater treatment takes place in what is, in effect, a special type of chemical plant. The installation and operation of these raise the cost of water service provision. Such costs are greater according to the range of pollutants dealt with and the degree of purification in respect of any single pollutant. Since the quality of water demanded by different users varies, an important issue is whether treatment should be to uniformly high standards or whether consumers with relatively low quality requirements, such as agriculture and certain industries, in contrast to households, should receive water treated to lower standards and consequently at less cost. This is the **dual supply** debate.

For example, modern treatment prior to the supply of drinking water to the public in the UK is concerned primarily with lead, nitrates, pesticides and **cryptosporidium**. A dual-quality supply system could reserve high quality treated water for essential domestic purposes and provide lower quality water for non-**potable** functions. Rees with Williams (1993: 39–40) argue that in the UK the bulk of water supplied conforms to the purity standards demanded by the EU's directive on drinking water. However, only 10–15 per cent of the demand for water requires such high standards.

Modern (or "advanced") treatment has been highlighted here. But this was not an issue before the 1970s. Traditional treatment consisted (and still consists) of sand filtration – to remove suspended particulate matter (such as dirt, dust, clays, rust, colloids, mineral matter) and algae, vegetable matter and bacteria; and chlorination – to kill any remaining bacteria and provide prophylaxis in the system in case of any colonization.

2.5 Distribution and consumption

One now turns to the distribution and consumption of water in Figure 2.1. In fact, consumption will be summarily dealt with here, because it is the focus of Chapter 4. On a global scale, the major users of abstracted water are agriculture, manufacturing industry, power generators, other municipal and commercial institutions, and domestic households.

There seems to be an important boundary question in respect of abstracted water distributed to users by pipes, canals and channels. When does a specific flow of water cease to fall under the consideration of supply and come under that of consumption? One definition of the boundary is where the pipes carrying water become the legal responsibility of the user of water. This definition encounters the difficulties that, for individual connections, legal responsibility may change; that for the same connections there may be joint legal responsibility, even if the consumer owns the pipes; and that, from a physical point of view, water flowing along a user's connection is clearly at that point being supplied, not consumed. The alternative definition is that consumption begins at the point where the water flow is put to use for **irrigation** and livestock, for cooling and washing in industry, and for the multiplicity of domestic uses such as cooking, washing, gardening and flush toilets. Where feasible, the second definition is used here. In any case, for either definition it is possible to write the global water balance equation:

water consumption = water production minus water losses   (2.1)

Water losses are never lost to the hydrological cycle, of course, and even within the hydrosocial cycle these flows may contribute in an unplanned way to plant growth in agriculture or to groundwater sources subject to later abstraction. The economic loss is the resources wasted in abstracting, treating and distributing the water prior to its leakage.

The distribution of water embraces both bulk transmission and retail networks to individual consumers, be they farmers, industrialists or families. No water grids exist comparable to those of electricity, because water is bulky, heavy and prohibitively expensive to shift uphill over long distances, although **gravity flow** systems can be very complex. Bulk flows from abstraction and storage locations almost always require pumping facilities. The more forceful the pumping action, the greater the water pressure and the faster the rate of flow. Pressure is defined as the force exerted per unit area at the base of a column of water of defined height; for example, 60 metres. Note that fresh water is usually pumped at pressure, whereas storm- and wastewater **sewers** operate under gravity.

By pooling risks, regionally integrated supply networks lower the overall need for safety-margin capacity. Where a shortage of demand exists in one region and a surplus in another, interregional transfers may be set up. These require major investment in reservoirs, pipes, pumping stations, and the usual machinery of valves, meters, and so on. In emergency conditions, water can be supplied by road tankers and it may also be transported by ship, as is the case in the Spanish islands. Exotic proposals to transport water as icebergs have not yet been realized. In urban areas, water for domestic households is commonly shifted by mains pipes, with an individual connection to each house or to each block of flats. Mains, comparable to mass transit systems in the field of transport, can be most impressive engineering structures. The London Water Ring Main, for example, is 80 km long and in 1996 supplied 1300 Ml per day of drinking water.

In many countries, the loss of water as a result of leakage is considered to be of great importance. This occurs between the points of abstraction and the points of consumption, as well as on the consumer's property. The significance of such losses is that, for a given level of consumption, the capacity of a water system in terms of abstraction, storage, freshwater treatment and distribution must be correspondingly higher to compensate for such leakages, with all the financial and environmental costs thereby implied. Losses are a function, in part, of the total distance over which water is distributed. For that reason they are often expressed in terms of cubic metres per day lost per kilometre of the mains supply. But burst pipes and the rate of leakage from them also vary directly with water pressure, which is higher, for example, in distribution to consumers in hilly areas and in densely populated areas. The latter also may suffer losses because their pipes are old. Losses are also exacerbated in districts where ground conditions are unstable or corrosive, leading to more frequent pipeline fractures.

Infrastructural renewal aimed at reducing losses takes several forms, such as leakage control targets defined in litres per property per hour, pressure reduction measures and the active location and repair both of bursts and old, corroded mains.

Loss control can be advantageous. For a given demand forecast, a lower leakage rate permits lower-capacity construction; for an existing system in which supply exceeds demand, it reduces operating costs in abstraction, treatment and distribution; for an existing system in which demand exceeds supply, it allows for more modest or no additions to capacity to meet that excess demand than would otherwise be the case.

To complete this review of freshwater distribution, it is worth noting that leakages are usually treated as a demand-side issue in current technical literature. For example, industry forecasts of the growth in demand include unaccounted-for water; lost water is treated as a use; and it is suggested that the control of distribution losses is a form of **demand management**. This does not seem helpful. A manufacturer of refined sugar, when considering losses from output because of pilfering, or contamination while in the warehouse or destruction in a road or rail accident en route to the supermarket, would never regard this as a demand for sugar, a bizarre act of consumption by a consumer whom the sugar never reaches. The manufacturer would regard all of these as storage or distribution losses prior to consumption.

So, the key argument for regarding **water losses** as part of supply analysis is that they occur prior to consumer use. The exact scale of leakage is not known, because of the imprecision of flow monitoring and metering, and this is why the term "unaccounted-for water" is to be found in the literature. However, it is believed that most losses occur in the mains supply, the distribution systems and company communication pipes, rather than from the supply pipes located on users' property.

The "supply or demand" question is not pedantic. To include losses on the consumption side misleadingly inflates the estimated demand for water. In countries where losses run at 30–50 per cent, or even up to 85 per cent in some cases, the overstatement of demand is enormous. It also deflects attention from the appropriate supply-side responses. If, for example, 50 per cent of water abstracted is lost between the points of abstraction and the points of use, then the reduction of these losses should be a prime option for action. This does not mean that loss reduction is cost free, nor that zero losses are either feasible or, in economic terms, desirable. What it does imply, to continue the sugar metaphor, is that raising consumption by means of increased delivery rates can rank equally alongside abstraction capacity increases as engineering and economic responses worthy of investigation.

2.6 Storm- and wastewater collection and treatment

Because of the physical character of the consumption process, some users of water use up most or all the water they receive and return little or none into the rivers and seas. Spray irrigation is an example. Other users exhaust little of the flows they receive and discharge the bulk after use, but the water they return may be degraded. Hydroelectric power stations, the textile industry and fish farming are examples here.

In some cases, most significantly in the industrial sector, consumers that do not exhaust all the water supplied to them are able to re-use that supply for themselves, over and over again, although water treatment may be required as part of this loop. The advantage to the consumer and to the environment is that such **internal re-use** necessitates less primary abstraction to meet consumption requirements.

Where water is not used up in the process of consumption, nor internally re-used, it leaves the consumption sector as wastewater. Most wastewater then requires collection for treatment in specially designed plants. Industry may also pre-treat its own wastewater. In the case of domestic wastewater, the collection of sewage is through pipes connected to a main sewer and thence to a sewage works. The components of the **sewerage** system collecting domestic and industrial waste-water are called **foul sewers**. In addition, there are **storm sewers** for the collection of rainwater, both to complement natural drainage at times of heavy precipitation and to deal with urban runoff. Rainfall draining from a housing estate also may be dealt with by a **soakaway**. Older towns may use combined systems of foul and storm sewers, which give problems of overload at sewage works during storms. In some cities, such as Canberra in Australia, stormwater is fed into artificial lakes, which are used both as settlement basins for cleaning stormwater before it is discharged and a source of water for consumption purposes.

Governments exhort sewage treatment works to use the best available techniques not entailing excessive cost (BATNEEC). Some of these wastes may then be externally re-used. For example, treated wastewater may be used for irrigation purposes. Internal and **external re-use** both contribute to the dual supply uses already referred to in §2.4 and both are to be found in Figure 2.1. Of course, external re-use may generate extra distribution costs; for example, in Canberra the sewage treatment works is many metres lower than the city itself, so re-use would require expensive pumping.

The nature of the treatment process is determined by the quality of the storm- and wastewater received at the sewage treatment works and the targeted quality of the effluent that leaves the works. Effluent limits may be decided on a case-by-case basis, depending on the river into which it is discharged and its use, as in the UK; or limits may be codified in general terms. Thus, "in Germany limits are fixed for all effluent discharges. In more than 50 annexes the law classifies industrial and community effluents and defines detailed limits for their discharges, either into surface water or the public drain and sewer system. The use of these limits is compulsory for all." (Heidebrecht & Hewitt 1994: 12)

2.7 Disposal

Wastewater that is neither internally nor externally re-used may be recycled, that is to say, released into the freshwater network, where it supplements the natural flows down stream from its point of disposal. A special case of **recycling** is the artificial recharge of aquifers already referred to.

Recycling suggests that a distinctive concept of the supply of water for abstraction is appropriate, which brings together both effective rainfall and recycling. Equation 2.2 states that the total supply of water available (S) for abstraction in a defined region or catchment area is equal to effective rainfall (E) plus the total volume of recycled water (I), all variables measured in m^3 per year.

$$S = E + I \qquad\qquad (2.2)$$

Equation 2.2 has the disadvantage of neglecting the location of a recycling point relative to its point of abstraction. Thus, recycling close to the headwaters of a catchment area would be counted as no better than the same volume recycled immediately above where a river flows into the salt sea, even though, in the latter case, such recycled water would bring virtually no benefit to either the riverine environment or the hydrosocial cycle in its totality.

Equation 2.3 states that the corrected supply for abstraction (S_c) is equal to effective rainfall plus the volume of recycled water corrected for the locations at which such water is discharged back into that region's freshwater system (I_c).

$$S_c = E + I_c \tag{2.3}$$

One practical way to apply a correction factor for any specific flow of recycled water per year at a defined point (let me call this flow I_p), would be to multiply that flow by a parameter δ_p, where δ_p is the distance by river of that discharge location from the sea divided by the distance by river from the sea where that flow was abstracted. This gives Equation (2.4):

$$I_{cp} = \delta_p I_p \tag{2.4}$$

In cases where the location both of abstracted water and its recycling into the freshwater system are the same, δ is equal to 1, and so the corrected value of the water recycled is the same as its absolute value. Where the discharge is straight into the sea, δ is zero, and so the corrected value of I is also zero. Where the discharge is up stream of its abstraction point, then δ is greater than 1 and the corrected value is higher than the absolute value. This approach reverses current practice of treating recycled water as some kind of negative need (a contradiction in terms) and places it squarely on the side of supply, with effective rainfall. Thus:

$$S_c = E + I_c = E + \sum_{p=1}^{n} \delta_p I_p \tag{2.5}$$

where 1 to n are the n points at which water is recycled.

In some countries, such as those of the EU, all discharges require government consent, whether they be into rivers, estuaries or the sea. If sewage **sludge** has not been re-used as a fertilizer or soil conditioner, it may be incinerated, buried in **landfill** sites or dumped into the sea. Rivers have the power to assimilate organic waste. This requires a minimum acceptable flow to make possible the process of effluent dilution and breakdown. Minimum acceptable flows are in any case necessary to protect riverine environments. The four major indicators of high overall river quality are high **dissolved oxygen**, low **biochemical oxygen demand** (BOD), low ammoniacal nitrogen and – other than in times of flood – low suspended solids.

2.8 Case study: a water balance statement for the Thames river basin

The preceding discussion suggests it is possible to create a water balance statement for management planning, which is generic to any catchment, region or country in the world. Such a statement would enable water resource managers to set out a full quantitative tabulation both of the supply sources of the hydrosocial cycle and the final users of those supplies. In any specific tabulation, the statement would be adapted to the particularities of the local situation.

Table 2.1 A water balance statement for catchment management planning.

Forms of supply	Average quantity (Ml/d)	Users	Average quantity (Ml/d)
Groundwater sources	A	Households	T
Surface water sources	B	Agriculture, forestry and fishing	U
Desalination of salt or brackish waters	C	Mining	V
Import of water from another catchment	D	Manufacturing	W
Internal re-use of waste water	F	Public services	X
External re-use of waste water	G	Commercial sectors	Y
Less: leakages and evaporation	H	Other users	Z
Less: export of water to another catchment	J		
Fall or rise in volume of stored water ($\pm$)	K		
Total net supply	A+B+C+D+F+G– H–J±K	Total use	T+U+V+W+X+Y +Z

Note: Total net supply equals total use

Table 2.1 provides such a statement. Column 1 refers to abstraction sources from the great wheel of the hydrological cycle and column 2 sets out the quantity of water supplied over the course of a year. The statement is designed to make explicit the circular components of the hydrosocial cycle, the scale of water losses and the import/export of water, so that the tabulation can assist in the development of sustainable water policies. The first supply row refers to abstractions from aquifers. In a specific application, planners may wish to specify the name, location and quality of these aquifers, and to distinguish deep-water mining from shallower sources. The second supply row indicates abstractions from rivers, lakes and estuaries, excluding water that will be desalinated. Row 3 refers to the abstraction of salt or brackish water, which is then desalinated. Rows 4 and 8 indicate, respectively, the import and export of abstracted water into and out of the catchment. Water exports are given a negative value, because they reduce the supplies available to home catchment users. Row 5 covers the *internal* re-use of wastewater. To take an example, a factory purchasing 10 Ml per day for industrial purposes from a pumped **groundwater** source, which then disposed of the same quantity as wastewater into a nearby lake, would be recorded as contributing nothing to internal re-use. Now suppose that environmental engineering works are carried out so that each day 10 Ml are still used for cooling, cleaning and manufacturing processes, but only 1 Ml is purchased, and 9 Ml are derived from the internal recovery and treatment of wastewater. In this second case, 1 Ml would be recorded for entry into row 1 and 9 Ml for entry into row 5. Row 6 refers to the *external* re-use of wastewater. The approach is the same as for row 5, except that the wastewater is not re-used within the same institution. Instead, it is transferred to another institution, possibly in another user group. Row 7 indicates leakage losses and evaporation at any point after abstraction and prior to arrival at the gate of the consumer. Like the export of water, it is set down as a negative value. Row 9 is set out only for those cases where the change over one year in the volume of water stored in reservoirs, water towers and underground tanks is not negligibly small in relation to total net supply. Where the change in volume is relatively large, say at least 1 per cent, a fall in stored quantity would be recorded as a positive value, and an increase as a negative value. The running down of stocks for a limited period of time permits greater consumption than would otherwise be possible.

Now let us consider columns 3 and 4. This is merely a classification of user groups, alongside the quantity consumed. Note that Table 2.1 does not deal with water losses after supplies have arrived at the user's gate, although it could readily be adapted to do so, if that were deemed appropriate. Note, too, that the volume used must include use made possible by the recovery of wastewater, that is from the forms of supply in rows 5 and 6. If data on re-use supply is not available, then the volumes of wastewater re-used should not be recorded. Finally, if all the calculations have been carried out correctly, total net supply would equal total use. This can be written as the identity:

$$A + B + C + D + F + G - H - J \pm K \equiv T + U + V + W + X + Y + Z \qquad (2.6)$$

The remainder of this section records the attempt to operationalize Table 2.1 in a specific catchment and from this to develop concepts of catchment self-sufficiency and stress.

In England and Wales, one of the responsibilities of the government's Environment Agency is these countries' eight water regions. Previously, they had been the charge of the National Rivers Authority, a body that was merged into the Agency in April 1996. The Thames Region (i.e. the Thames river catchment: Fig. 2.2) stretches from Wiltshire in the west to the Thames **Estuary** in the east, from London in the north to the Surrey Downs in the south, an area of nearly 13 000 km². The

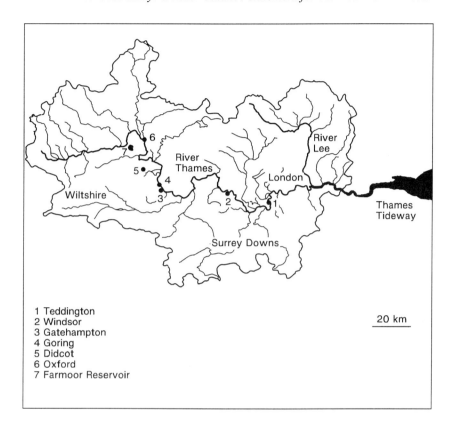

Figure 2.2 The Thames catchment

region has a population of 11.5 million and is one of the most intensively used water resource systems in the world. It is dependent in the main on only one river, the Thames itself, which at Teddington Lock in southwest London has an average flow of 5667 Ml per day. Average annual rainfall is 704 mm, based on the standard 1941–70 period. The effective average annual rainfall available to sustain river flows and replenish natural underground storage, is 250 mm. The 1:50 year effective drought rainfall is 100 mm. Rainfall of 1 mm is equal to 13 million m^3. Winter rainfall is at least four times greater than summer rainfall. A substantial area of the region is composed of permeable water-bearing strata, and the major aquifers are chalk, Jurassic limestone, Lower Greensand, and river gravels (NRA Thames Region 1994, NRA 1995).

The water balance statement for the Thames Region is set out in Table 2.2. Let us consider first the forms of supply, which represent actual abstracted quantities for 1994. Groundwater sources provided 1569 Ml per day, 30 per cent of the known total gross supply of the first six rows. The aquifer sources are replenished mainly in the winter. There are well over 300 boreholes in daily use in the region. Licence

Table 2.2 Thames region water balance statement for 1994.

Forms of supply	Average quantity (Ml/d)	Users	Average quantity (Ml/d)
Ground water sources	1569	Households	2091
Surface water sources	3447	Agriculture, forestry, fishing	417
Desalination of salt or brackish waters	0	Mineral washing	56
Import of water from another catchment	137	Manufacturing (direct abstraction)	535
Internal re-use of waste water	not known	Public water supply operational use	24
External re-use of waste water	not known	Other uses	1083
Less: leakages & evaporation	−856		
Less: export of water to another catchment	−91		
Fall or rise in volume of stored water (±)	negligible		
Total net supply	4206	Total use	4206

Note: Total net supply equals total use.

holders with the largest abstraction permissions for ground- and surface water are six statutory water companies, of which by far the largest is Thames Water Utilities Ltd, and some industrial companies. Water supplied by the statutory companies is known as the public water supply.

Surface water sources in 1994 supplied 3447 Ml per day, 67 per cent of total gross supply. The bulk of this comes from the Thames. The majority of abstraction in volumetric terms is between Windsor and Teddington, but there are also significant withdrawals at Farmoor near Oxford and on the lower River Lee. Post-abstraction, the London supply system relies on the large storage reservoirs of the Lower Thames and Lee Valley (NRA Thames Region 1994: 3, 58).

Desalination is not a source of fresh water in the region. Imports are only 3 per cent of the known gross total supply. Internal and external re-use exists, but the quantities are not known. The NRA states (1994: 37):

The value of re-use has already been realized in many sectors of industry and commerce. Many large users/dischargers have recognized the cost savings from (re-using) water, not necessarily from reducing the costs of treated water, but from the savings of reduced trade effluent disposal. At a smaller scale, many vehicle washes in the region now (re-use) much of the water used.

Leakage (and evaporation) is equal to 17 per cent lost to gross supply. An important arithmetic point arises here. I here calculate leakage as the ratio of losses to the sum of ground abstractions, surface abstractions and imports. Leakages are not taken as a percentage of net supply, to avoid losses appearing both in the numerator and denominator of the expression. This simple point is worth clarifying in more general terms. Suppose in Table 2.1 the only supply-side entries were *A*, *B* and *H*. My presentation of proportionate leakage losses (let us call this Γ) would be:

$$\Gamma = \frac{H}{A + B} \tag{2.7}$$

However, if anyone were foolish enough to present leakage losses proportionate to total net supply we should have:

$$\Gamma = \frac{H}{A + B - H} \tag{2.8}$$

and no meaning whatsoever attaches to this expression. For precisely the same reason, given that total net supply is equal to total use, it would be nonsense to calculate leakage losses as a proportion of Table 2.1's total use figure.

The NRA uses the term "total treated water losses" to refer to leakage losses from trunk mains, distribution systems and service pipes. The third of these, service pipes, consist of the company's communication pipe and the consumer's supply pipe. Note that leakage losses from the Thames Water Utilities public water supply was as high as 29 per cent in 1994–5. (Ofwat 1995: 21)

Finally, with reference to the sources of supply, note that the export of water is even smaller in scale than water imports from other catchments. These exports go to the Anglian Region – to the north and east of the Thames Region. As far as change of volume in reservoirs is concerned, these were negligible in 1994.

Now let us look at the user categories. Households are the largest single consumer group, with 50 per cent of total use. Freshwater sources both from the statutory companies and private water supplies are included. In the home, about 150 litres pppd are used, the bulk for flushing toilets, for baths and showers, and for washing machine use. Garden watering use is modest, except in hot, dry summers. In 1996, fewer than 5 per cent of households in the Thames Region paid by meter (NRA Thames Region 1994: 10).

Consumption in agriculture, forestry and fishing is 10 per cent of the total. Most of this is for growing water-cress and for fish farming. Farmers can be big users in hot weather, with spray irrigation. Direct abstractions (i.e. abstractions by private persons or institutions other than the statutory companies) are the main source of water for agriculture.

Mineral washing uses 1 per cent of total net supply and the operational use for the public water supply is even less. Manufacturing sourced by direct abstractions consumes 13 per cent of the total. Finally, "other users", at 26 per cent of the grand total, refers to any users, other than households and agriculture, that take their water from the public water supply.

In a nutshell, households are responsible for 50 per cent of total use, agriculture for 10 per cent and the residual category, "industry", takes 40 per cent. Industry embraces power generation, sand and gravel washing, brewing, other manufacturing, commerce and the public sector. About 72 per cent of industrial use is metered. There is a large power station on the non-tidal Thames, at Didcot, which abstracts water for cooling. But the power stations on the tidal river, which also abstract water for cooling purposes, have no significant implications for freshwater resource management (NRA Thames Region 1994: 7–8, 22).

To end this case study, I shall assess the self-sufficiency of the Thames Region in its water resources and introduce a measure of quantitative stress on the catchment's water resources. I shall define self-sufficiency in any year in any catchment (or region or country) as a situation where effective rainfall plus recycling is at least equal to total use. Recycling would be measured in its corrected form (see Eq. 2.5) where that is available, otherwise in the non-corrected form. To facilitate inter-catchment comparisons, all the data will be presented in terms of millimetres per year.

So, the measure of self-sufficiency requires information on just three variables in any given year: effective rainfall, total use and the volume of recycled water. In the case of the Thames Region, the first two are readily available. Effective rainfall in 1994 was 268mm. In the same year, Table 2.2 shows total use was 4206Ml per day, equivalent to 118mm. But what about the volume of recycling?

Hydrologists and engineers traditionally distinguish between **consumptive** and **non-consumptive** uses of water. In any particular activity, the greater the proportion of the water used which is then lost to the catchment's fresh ground- and surface waters, the more consumptive it is. Such losses occur either because of **evapotranspiration** during use or because water is embodied in the product itself, as in brewing. In brief, the less the volume of water recyclable, the more consumptive is the use.

In the Thames Region, some non-consumptive activities exist. Fish farming and water-cress production are among these, in contrast to spray irrigation. Power generation stands in the middle of the spectrum.

National Power operates a major power station at Didcot on the middle Thames where approximately 142Ml per day on average can be abstracted for use as cooling water, the licence requiring between 50 to 66 per cent of that abstracted, depending on prevailing flow conditions, to be returned to the river after use. (NRA Thames Region 1994: 28)

With respect to abstraction for the public water supply to households and industry, several hundred sewage treatment works recycle their treated effluent to the River Thames. The exception is London, where water abstracted for supply is in the main returned as treated effluent to the Thames tideway, on average 2400Ml per day.

As the NRA Thames Region states (1994: 36):

The (recycling) of effluent is already widely practised within the Thames Region by virtue of the geography of the River Thames with by far the major-ity of surface water abstraction occurring at its downstream freshwater limits. The security of supplies for London relies on water abstracted up stream being returned after use as good quality treated effluent. Good quality effluent is a valuable resource and this practice of abstraction, use and (recycling) must continue, relying on the treatment of effluents to the highest standards. In recent years, the NRA has advocated that water abstracted should be used and returned as treated effluent up stream of the original abstraction. This approach was used to licence the Gatehampton source (near Goring-on-Thames) which will be mainly used up stream to supply the Oxford area, the effluent being returned to the River Thames via Didcot and Oxford sewage treatment works.

As far as I am aware, the NRA has never explicitly estimated for any of its regions the total volume of water recycled to the catchments' fresh-, ground- and surface waters. This is a serious omission in its information base. However, in the Thames Region an Environment Agency source has suggested a value of 20–25 mm, excluding London's wastewater discharge to tide.

So, we have a total resource available, for the self-sufficiency calculation, of 268 mm from effective rainfall and 25 mm available from recycling, against total use of 118 mm, a ratio of supply to use of 2.48. The Thames Region in 1994 was clearly self-sufficient in its water resources.

Finally, I wish to develop a catchment stress indicator. The simple argument here is that, the greater the level of abstraction within a catchment in relation to effective rainfall, the greater is the stress imposed on the river basin as an ecosys-tem. But such stress is reduced by system leakages and recycling. So, we have:

$$Q = \frac{(A + B) - (H + I)}{E} \tag{2.9}$$

where Q is the quantitative measure of catchment stress, A is groundwater abstraction in the basin, B is surface water abstraction in the basin (but excluding tidal waters), H is leakage, but excludes evaporation, I is the uncorrected value of the recycled flows, and E is effective rainfall. All of these terms are most usefully presented in millimetres for the catchment area.

In the case of the Thames Region, using the data already presented, we have:

$$Q = \frac{(44 + 97) - (24 + 25)}{268} = 34\% \tag{2.10}$$

If other researchers can carry through a similar calculation, intercatchment comparison will be interesting. The stress value in the Thames Region is aggra-vated by the density of the basin's working and residential populations, but reduced by its humidity and the considerable scale of recycling. Hydrologists may also wish to assess whether leakage is an offset to quantitative stress, as implied by the *H* term in Equation 2.9. For example, in dry years in the Thames catchment, leakage from a pipe buried at 1–1.5 m is likely to move vertically towards the surface rather than downwards to recharge any depleted water table.

2.9 Case study: the hydrosocial cycle in Bratislava

The Slovak Republic lies in Central Europe, bounded by Poland, the Ukraine, Hungary, Austria and the Czech Republic (Fig. 2.3). Its geographical area is 49000 km^2 and in 1995 its population totalled 5.4 million, a density of 110 persons per km^2. The Slav peoples entered the present area of Slovakia in the fifth century AD, but it is only since 1 January 1993 that the country has achieved a stable independence, following the 1989 Velvet Revolution against Soviet hegemony and the subsequent break-up of Czechoslovakia. In May 1995 the rate of exchange of the US dollar to the Slovak Crown was 1:28.

The River Danube forms the southwestern border of the country and it is on this river, down stream from Vienna and up stream from Budapest, that Slovakia's capital is to be found, Greater Bratislava. In 1995, the city's population was some 500000 persons.

The Danube is the second longest river in Europe, at 2850 km. At Bratislava, the flow averages 2060 m^3 per second, but varies from rates as low as 570 to those as

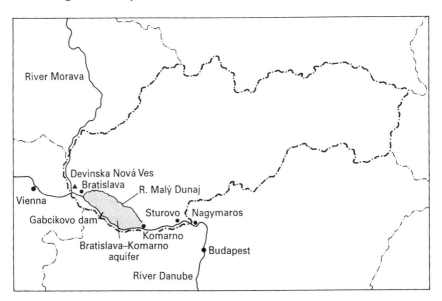

Figure 2.3 The Slovak Republic.

high as 10600. The flow rate is highest usually in May–June, when thawing snow from the Alps feeds the river up stream of the capital.

However, with so many urban areas located above Bratislava, the Danube is polluted by the time it leaves Austria and is joined by the Morava from the Czech Republic just to the west of the capital. In an interview at the Ministry of the Environment, I was told that the pollutants are sewage discharge pathogens, nitrates from fertilizer use, heavy metals, mineral oil discharges, chlorinated solvents and acid rain.

In fact, Bratislava has a second choice for its drinking water supplies and this is the aquifer, said to be the largest in Europe, which begins in the area of the capital and runs as far as Komárno, 80 km to the southeast. This aquifer lies between the Danube and the Malý Dunaj River to its north. It is from this source that Bratislava abstracts the major part of its household and industrial needs, on a scale of some 83 million m^3 per year. Abstraction is licensed under the Water Act 138/1973.

In 1995 the Ministry of the Environment was the regulator responsible for protection of the water environment, and the Ministry of Agriculture dealt with water utility functions, hydroelectric power and irrigation. Slovakia has four catchment-area bodies, one of which is for the Danube and has its headquarters in Bratislava. The Ministry of the Environment produces hydroecological plans, and the Ministry of Agriculture prepares water management plans, but it is in fact the same extra-ministerial technostructure that writes both sets of plans, principally from the Water Research Institute, a state-sector research organization of some 300 employees.

In 1995 the capital's fresh- and wastewater utility was a state-owned company, Vodárne a Kanalizacie Bratislava. After the rise to power of the Communist Party in 1948, a national system of planning for water supplies had been laid down in a series of five-year plans. Within this context, water companies had been nationalized. In the early 1990s, following the end of Communist power signalled by the Velvet Revolution, discussions had begun on the privatization of the industry. In 1995, the intention was to shift from a wholly state-owned company to one in which shares are held. These shares were to be divided as follows: 35 per cent by the local municipality, 34 per cent by central government, 28 per cent through voucher privatization, and 3 per cent through property restitutions.

There is no general system of dual supply of water, so the bulk of customers receive water of potable standard, whether they require this or not. The water quality of the aquifer is high and, with one exception, water from the boreholes is only chlorinated, no other treatment being regarded as necessary. The water conforms to standards equivalent to those required in the EU's drinking water directives.

As already indicated, water is imported into the city by pipeline from boreholes located in the area in and around Greater Bratislava. There are no surface reservoirs, but 27 underground tanks exist with a total capacity of 229 million m^3. Distribution is by mains pipes and individual connections to houses and blocks of flats. Losses occur in the mains supply and the individual connections and are ascribed mainly to the ageing of the distribution network. Losses are claimed to be only 14 per cent of total treated fresh water. Measures taken to reduce this include leakage control targets, pressure reductions and the repair and replacement of old mains. Of the 1995 consumption total, 58 per cent was for domestic use, 19 per cent for industrial use and 23 per cent was other uses.

With regard to wastewater, the Ministry of the Environment believes the scale of re-use within firms is negligible. Nor is any treated water re-used by other consumers. So, the internal and external re-use loops of the hydrosocial cycle, discussed earlier in this chapter, hardly exist. Industrial **pre-treatment** of water does take place, although no data existed on its scale. The foul- and storm-water systems are combined, and overload problems have always existed in the past. However, in 1993 a major programme of placing water meters in domestic homes was undertaken, for example on the huge Petrzalka housing estate with its 140 000 population living in high-rise blocks. There has also been a real increase in fresh- and wastewater charges. As a result, consumption has been reduced and there have been no overload problems in the recent past. The capacity of the three wastewater treatment plants, at Petrzalka, Devinska Nova Ves and Vrakuna, is equivalent to that required for 1.5 million people and is therefore sufficient for the capital's households and its industry, hospitals, schools, and so on. The treatment processes are mechanical–biological, with additional treatment of the sewage sludge. One hundred per cent of wastewater is treated. Vrakuna is about 15 years old, whereas the other two plants are very recently completed. The biological oxygen demand of wastewater released into the river is 10 mg per litre, well below the requirement set by central government of 17 mg per litre.

Wastewater is released into the Danube, the Malý Dunaj and the Mlaka rivers. Discharges require government consent, under the 1973 Water Act, and they are made both by the treatment plant itself and by firms. The sewage sludge is stored at the wastewater treatment plants. This poses problems for the future. Various proposals exist for the use of sludge in agriculture and in forestry, but great care will be necessary to ensure that the chemical composition of sludge is appropriate to the different varieties of Slovak soil.

The most important single user of water for industrial purposes in Bratislava is Slovnaft, the state-owned refinery and petrochemicals complex located at the southeast corner of the city on the Danube's north bank. In terms of the land area it occupies, this enterprise is the largest in Slovakia.

Slovnaft was built on the river for ease of access to cooling water, water for other processes and as a release point for the plant's wastewater. Abstraction from the Danube is of the order of 79 million m^3. As can be seen, this is almost as great as water production from aquifer sources for the whole of the city. Part of the enterprise's wastewater is pumped untreated straight back into a branch of the Danube. Part is returned after treatment, but information on the ratio of these two flows was not available to me.

The wastewater treatment plant combines mechanical, chemical and biological technologies, and went into operation first in 1986. The quality of water leaving the plant is said to be comparable with the water abstracted from the Danube. Indeed, it is claimed to be of higher quality in respect of suspended solids (15 mg per litre against the river's 34 mg per litre) and hydrocarbons (1 mg per litre against up to 200 mg per litre for the river). The annual costs of the sewage treatment works were 70 million Slovak crowns in 1994.

The increasing real cost of water consumption has led Slovnaft's management to consider internal re-use, and an estimate has been made that 100 per cent re-use would require an investment programme of SKR1200 million. Plans are also being made to raise the capacity utilization rate of the sewage treatment works as well as increasing its theoretical capacity, so that the discharge of untreated wastewater could end by the year 2000.

Problems exist with leakage from the plant's 400 oil and petrochemicals storage tanks, contaminating the soil and groundwater. Slovnaft is sited at the eastern point of the Bratislava–Komárno aquifer. New storage tanks will be required to end this pollution. At the same time, confusingly, Slovnaft's Environmental Department head, Mr Juraj Lalik, stated in 1994 that the company's groundwater protection unit built by Geo-test of Brno, is 100 per cent efficient. In 1971, parts of Bratislava were without drinking water for up to six months because Slovnaft had contaminated the aquifer close to the city. Since that time, a **hydraulic** wall has been put in place to prevent polluting **infiltration**. After the 1971 incident, a borehole was sunk at Kalinkovo (30 km from Bratislava) and this is still used in addition to the wells sunk in the immediate Greater Bratislava area. Kalinkovo exists as a reserve source and also as a means of checking groundwater quality in the area of the Gabcíkovo dam. Two hundred litres per second were pumped in 1995 (Krejci 1994).

Abstraction on a smaller scale by specific industrial firms exists, both from the Danube, the Malý Dunaj and the aquifer. Istrochem is an example, the giant chemical plant located in the northeast of Greater Bratislava. These direct abstractions by Slovnaft and others are not included in the total of 83 million m^3 referred to above, which is the output of Vodárne a Kanalizacie Bratislava from the aquifer.

Before ending this case study, it is appropriate to consider briefly the construction and function of the Gabcíkovo dam, which is located 50 km down stream of Bratislava. The preparation of a common strategy for the exploitation of the Danube by the Czechoslovak and Hungarian authorities first began in 1952. The principal objectives were threefold: to create a hydroelectric power source, to control flooding, and to improve the river's navigability. In September 1977, the governments of Czechoslovakia and Hungary signed an agreement to begin construction of a dual barrage system at Gabcíkovo in Slovakia and at Nagymaros in Hungary, 30 km up stream from Budapest. Work began in the following year.

However, in both countries, the project met strong opposition from environmentalist groups. It should be remembered that in eastern Europe, before the disintegration of the USSR during the Gorbachev era, environmentalism was an important mode of organization and protest for a wide variety of groups opposed to the dominance of the ruling communist parties and the hegemony of the Soviet Union. As a result of widespread opposition in Hungary, the government unilaterally withdrew from the twin barrage plan in 1989. In 1991 the Czechoslovak government decided to proceed alone, with works which could be carried through unilaterally. These consisted of the Cunovo–Hrušov lake, a network of dykes, channels, off-take structures and dams, a navigation channel, and at Gabcíkovo itself, a dam, the power plant and two locks for shipping.

The dam was completed in October 1992. It is clear that the river/aquifer infiltration regime has changed, but defenders of the project quickly claimed that environmentalist predictions of a fall in groundwater levels and a deterioration of the quality of groundwater had failed to materialize. They suggested that, in fact, the megaproject had increased the flow of water in branches of the Danube, and their associated lakes, particularly the 150km Malý Dunaj branch, which was a living river again. At the same time, the flow of water in the old Danube river bed, on the Hungarian side, had virtually ceased. Moreover, the failure to carry through the Hungarian side of the project had resulted in navigation difficulties for large vessels around Štúrovo and Nagymaros. In the year 1994, the hydropower plant produced 2200 million kW hours of electricity (Anonymous 1995, Binder 1993a,b).

Gabcíkovo–Nagymaros can be seen as an engineering dream to tame the waters of one of Europe's greatest rivers and to divert them to technocentric objectives, at the cost of huge environmental disruption. It is difficult to believe it could happen again in an era when the absolute dominance of engineers within a "democratic-centralist" society no longer exists (Gindlova et al. 1995).

2.10 Final remarks

This chapter contrasts the interdependent relationships between the hydrological cycle and what I have called the hydrosocial cycle. The former is a full cycle, whereas the latter is only a partial cycle, through the recycling and internal and external re-use loops. These loops are of great relevance to the sustainability agenda, and we return to this issue in Chapter 7.

This chapter shows how varied are the options available for increasing the supply of fresh water to agricultural, industrial and domestic consumers. There are 13 such options:
- the erection of new infrastructures to abstract fresh water from sources below the ground or from those on the land's surface
- abstraction from estuaries and the sea
- the expansion of existing abstraction plant, to raise installed capacity
- infrastructural works to raise usable capacity closer to theoretical capacity
- the reassignment of authorizations to abstract between licensees, increasing the rate of abstraction within a single catchment area
- more sophisticated control in the management of existing reservoirs
- more effective **conjunctive use** of rivers, underground and reservoir sources
- the expansion of the storage network
- investment and changed management practices to reduce water losses from

existing supply networks
- interregional transfers, where a shortage of demand with respect to supply exists in one region and a surplus in another
- the extension of the internal re-use of water
- the extension of the external re-use of water
- the provision of infrastructures to extend recycling to raise the measure of the "corrected supply" available for abstraction.

In addition to these 13 options to raise the quantity of water supplied, proposals exist for the development of a dual supply system, in order to reduce water treatment costs.

Finally, the first case study develops and applies the concept of a water balance statement, as well as measures of self-sufficiency and quantitative stress at the catchment level. The second case study indicates something of the complexity of the hydrosocial cycle in a large city.

CHAPTER THREE
Supply: the economist's perspective

3.1 Capital and current spending

Chapter 2 has given a brief account of the nature of the hydrological cycle and a fuller description of what may be fairly called the hydrosocial cycle. The purpose of Chapter 3 is to consider the supply of fresh- and wastewater services as an economic phenomenon, principally with reference to the costs of production, although here excluding environmental costs, a subject to be returned to in Chapters 7 and 8.

The hydrosocial cycle illustrated in Figure 2.1 (see p. 6) draws on an immense variety of the economy's real resources. These include fresh water itself, at the point of abstraction; the land sites required for the system's infrastructure and a reserved catchment area, where land access and use may need to be restricted; boreholes; pumping stations; reservoirs; plant for mechanical, biological and chemical treatment of both fresh- and wastewater; an extensive system of pipes and other channels, including **aqueducts**, water mains, **culverts**, stormwater drains, sewers and individual house connections; a wide variety of monitoring, measurement and control devices, much of which may be computer-based; **switchgear**, and standby equipment to generate electricity in case of power failure; electric power itself; materials such as filter sand and chlorine gas; spare parts; the means of transport of sewage sludge and the use of landfill sites and aquatic media for disposal; and the human resources necessary to design, build, operate, monitor, maintain, repair, rehabilitate, manage and administer the whole. **Headworks costs** are those of abstraction, storage and treatment. **Network costs** are those of freshwater distribution, stormwater and **foul water** collection, and wastewater disposal after treatment (see Fig. 2.1).

In the analysis of the supply of water services, the economist converts the tabulation of real resources into money expenditures, on the basis of the resources' market prices. Expenditure can then be divided into **capital account spending** and **current account spending**.

Capital expenditure is defined as that where the resource purchased has an expected life of more than 12 months. This would include land; civil engineering infrastructures such as boreholes, reservoirs, works road networks, pumps, pipes and sewers; buildings and plant of all kinds; and vehicles.

Current expenditure refers to purchased resources that are either immediately used up in the process of production or which have a life of 12 months or less. These include fresh water itself, materials, chemicals, spare parts, electric power and labour time. In practice, relatively small expenditures, such as those on some types of office equipment, would be counted as current expenditure, even where their life exceeds one year.

Where long-life resources are rented rather than purchased, or in other instances where payment is made on a regular basis rather than as a lump sum, the expenditure will be classified as a current account item, on the basis that a company can withdraw from commitment to such a purchase, if it so decides, in the **short term**. So, in this particular case, the rent of land and buildings, and landfill payments, for example, are all likely to appear under current expenditure (Shute 1994: 51).

From an economic and a financial perspective, the importance of the distinction between capital and current spending is: investment in a newly developed system's civil engineering infrastructures and plant requires far greater expenditure on capital account than the current account costs of a single year's operations and, as a result, capital spending often requires debt finance; commitment to current account spending is far more flexible than the sunk costs under capital account; the balance between capital and current spending is of the greatest importance in project evaluation.

Chapter 2 concluded by setting out 13 options for increasing the supply of fresh water to agricultural, industrial and domestic consumers. It is possible to set out the relative proportion of capital and current expenditure within the total expenditure of any single option. Some options, such as the erection of new infrastructures to abstract fresh water from sources below the ground or from those on the land's surface would require relatively high capital spending. Others, such as more effective conjunctive use of rivers, underground and reservoir sources, may require relatively high current spending.

3.2 Abstraction charges

This section addresses a current expenditure unique to hydroeconomics, that is, abstraction charges. In every country in the world, rights exist to abstract water from ground or surface sources and such rights are defined in law or by customary practice. Where land is privately owned, these rights are usually enjoyed by the landowner, often circumscribed by government licensing. Abstraction rights can be an important attribute of land, reflected in its market price or its rental value. As such, the matter is traditionally handled in political economy by the application of David Ricardo's **theory of differential rent** (Ricardo 1970).

Abstraction charges are levied in two distinct cases. First, where a landowner, either with riparian rights or rights of access to groundwater, rents them out to an abstractor; secondly, where a state authority enjoys licensing rights over surface water or groundwater, and imposes charges on abstractors who wish to secure such a licence. From the point of view of economic theory, both cases are examples of rental payments, in the sense that both are charges for a scarce commodity which is a natural resource. In the first case, the rent the landowner receives is likely to be influenced by historical and cultural factors and, economically, both by competition between landowners in supplying abstraction rights and through the value that users place on access to water. The Australian study at the end of this chapter deals with **tradable abstraction rights**. In the second case, the rent accrues to the monopoly power of the state authority, which might charge little or nothing, or which might push the rent as high as the market will bear.

Schiffler and his colleagues at the Deutsches Institut für Entwicklungspolitik (German Development Institute) note that abstraction charges levied by the state exist in several countries, such as England and Wales, France, Germany and parts of the USA. In 1994, Jordan became one of the first developing countries in the world to introduce groundwater abstraction charges, but applicable only to industry (Schiffler et al. 1994: 13–16).

State abstraction charges may be levied primarily as a **fiscal** device, to raise money to cover government expenditures of one kind or another. For example, in the case of England and Wales since the mid-1960s, major abstractors have been required to pay charges based on the total quantity of water they are licensed to take. These charges are modest, currently covering only the operating costs of the riverine functions of the Environment Agency. At the same time, the system of abstraction licensing safeguards the investments of the water industry. Charges may also have an environmental function, restraining the volume of water abstracted so as to reduce stress on the hydrological cycle. However, in England and Wales, the level of charges is so modest that the Environment Agency believes they are too weak to shape consumer behaviour.

Charges may be in the form of a fixed payment, invariant with the quantity abstracted; or there may be a price which is constant, whatever the quantity taken up, as with the rate of 100 fils[1] per m^3 in the case of Jordan; or unit price may rise with the quantity abstracted. Moreover, allowances may be built into the pricing system, so that only a fixed charge is levied on a minimum abstraction level. One may also charge on the basis of the quality of the water withdrawn; in parts of Germany, higher charges exist on abstraction from cleaner and deeper aquifers and lower charges on shallower and more polluted ones. Charges may also vary with the season, as in France, where they are higher during the summer when water is scarce. Finally, abstraction charges may vary between consumptive and non-consumptive uses, being higher for the former. This is the case in England and Wales (Kinnersley 1994: 16–17, 44–5).

---

1. 1000 fils = 1 Jordanian dinar.

3.3 Prime costs and overhead costs

The discussion of capital and current costs in §3.1 helps prepare the way to consider the most important supply-side concepts in the industrial economist's tool-kit – the cost functions. These quantitative relationships describe the cost of supplying output in any time-period at each scale of output from zero units up to the system's theoretical capacity. In the analysis of the water industry, supply can refer, as appropriate, to any single stage in the hydrosocial cycle or to the system as a whole. In the text below, an example from freshwater treatment is used.

The ability to shift from one output level of treated fresh water up to a higher level clearly depends on the time available to make the change. It was Alfred Marshall, one of the greatest British economists, who introduced periodization into the analysis of industrial supply and demand. He distinguished between the short term and the **long term** (Marshall 1962). In its application to the hydrosocial cycle, this simple, twofold distinction can be represented in the following way: in the short term, increased daily output is possible through operational changes or by organizational innovations demanding new procedures, both of which are relatively straightforward in their introduction; in the long term, additional capital equipment is required either in new projects or for the expansion of existing infrastructure and plant, with associated financial, resource and organizational planning.

In conventional analysis it is possible, for either the short or long term, to calculate the total cost of supply at each level of output per day (or per month or per year). Average cost at each level of output can be derived from this by dividing the total cost by the number of units produced. The **marginal cost** of the nth unit of output is defined as the difference in total cost between producing n and $(n-1)$ units. Similarly, **marginal revenue** for the nth unit of output is defined as the difference in total sales income derived from selling n rather than $(n-1)$ units. This may be a positive or negative datum. The suggestion of neoclassical analysis is then that the management of a plant will set its output at the profit-maximizing level where marginal cost is rising, marginal revenue is falling, and the two are precisely equal.

A related but alternative exposition is set out below. It has the advantage that the terms refer to engineering categories, making it easier for non-economists to grasp quickly. It is also more appropriate than the conventional analysis, since it is a better representation of the way industrialists actually think about profitability, embracing the routines of what has been called "satisficing" rather than "optimizing" behaviour (N. Kay 1994, Nelson 1994).

Expenditures are defined using three categories:
- **Prime costs** are defined as those used up in the daily production of goods and services, and consist of the salaries and wages of the workforce, the costs of power, materials, spare parts, and other consumables such as bought-in specialist inputs and services (Amin 1994: 89).
- **Overhead costs** of production are non-prime costs and are often assumed to be fixed whatever the level of production. In the simplest case, where capital

expenditure on land, infrastructure and plant has been funded 100 per cent by loans, this element of annual overhead costs would be set equal to the annual interest payable and the **principal** repayable on the debt incurred. Where capital investment is funded from company profits, overhead costs are the **amortization** charges necessary in the long run to replace the plant when it is scrapped. To this fixed capital charge must be added the fees, salaries and other emoluments of the managerial élite, that is, the board of directors, without whom the company or corporation could not be said to have a legal existence. The rent of land and buildings, and the leasing costs of capital equipment such as vehicles, are also included in overheads.

• **Total costs** per year are equal to prime plus overhead costs.

For each level of output, one can calculate prime, overhead and total costs per unit of output, and these functions, called **average prime cost**, **average overhead cost** and **average total cost** respectively, are represented for a hypothetical case in Figure 3.1, illustrating the short-term situation. Average overhead costs fall (at a decelerating rate) as the unchanged level of cost is divided by ever greater levels of production. Average prime costs decline in a similar way because, in freshwater treatment, the prime resources required to produce higher output levels are not sharply greater than those for lower output levels – there is a "stickiness" in prime costs akin to those of overhead costs. However, as the plant moves closer to its theoretical capacity, average prime costs may begin to rise, indicating technical difficulties in achieving 100 per cent capacity utilization. Average total cost is simply the sum of the other two functions and so reflects their combined shape.

In Figure 3.1 the assumed price of output is set at the horizontal line P–P. (Explanation of the determinants of unit price is left to Ch. 4.) The difference between unit price and average prime cost at any specific output level is called the **gross margin** and the ratio of gross margin to unit price is a measure of the degree of monopoly, defined as "the conditions in a market which permit a particular level of the ratio of gross margins to be realized." (Robinson & Eatwell 1974: 155) These conditions, for industry in general, include the degree of business concentration, conditions of entry into a production sector, the level of product differentiation, and collusion between firms (Sawyer 1994: 433).

The difference between unit price and average total cost is the **net profit** margin. At 80 units of output, AC is the gross margin, AB the net profit margin and total net profit equals (80. AB). Net profits, after tax, are available for distribution to shareholders, in the case of a private company, and to funding new investment. It would be imprudent for a company to distribute the entirety of net profit to its shareholders, since net investment determines the dynamics of the institution. Similarly, when the water company is publicly owned, if all its net profits are appropriated by the local authority or by central government, its capacity to innovate technologically will be stifled.

The rate of profit on capital is the ratio of net profit to the value of capital. It can be expressed in *ex ante* or in *ex post* terms. The *ex ante* rate of profit is an estimate made of what the rate will be in future on a new investment; the *ex post* rate of profit states what the rate is on an investment already made. Project evaluation is the subject of Chapter 5.

In the context of water resource planning for the long term, an important concept is the minimum point of the average total cost curve (ATCmin) for new investments, either for the whole of the supply phases of the hydrosocial cycle or for specific parts of it. In Figure 3.1, this has a value of 9.9 at an output level of 100 units of output per day.

For example, in 1994 the European Commission decided to finance a comprehensive feasibility study for a water and sanitation investment plan for the city of Daugavpils in Latvia. Firms tendering for that contract were required to include an economist in their team and one of the responsibilities of the economist was to "determine the marginal cost (average incremental cost approach) of water and sewerage". The reference here is – rather confusingly – to the ATCmin for the case of new capital expenditure.

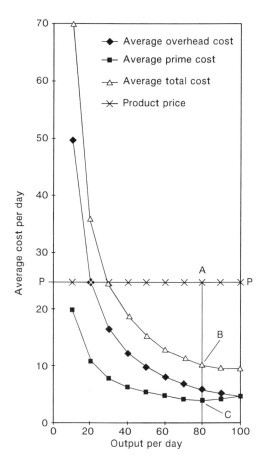

Figure 3.1   Short term supply: overhead, prime and total cost per unit output per day.

Some of the key relationships described in this section are illustrated in Figure 3.2. Output price can be divided into two components: prime costs per unit output and the gross margin. Prime costs are made up of wages, salaries, and the expenditure on power, materials, spare parts and other consumables. The gross margin covers overheads and is the source of net profit. Overheads are made up of the rent of land and buildings, any leasing costs incurred on capital equipment and vehicles, payments to the board of directors, interest payable on loans, and the amortization of the capital stock. Amortization is itself separable into the repayment of the principal of loans, plus any additional amortization allocations. Net profit is the source of taxes payable to government, **dividends** payable to shareholders, and of **retained profits**. Amortization, net of principal repayments, plus retained profit, are the internal funding sources for new rounds of investment, that is to say, for the **accumulation of capital**. These enterprise finance issues are returned to in Chapter 6.

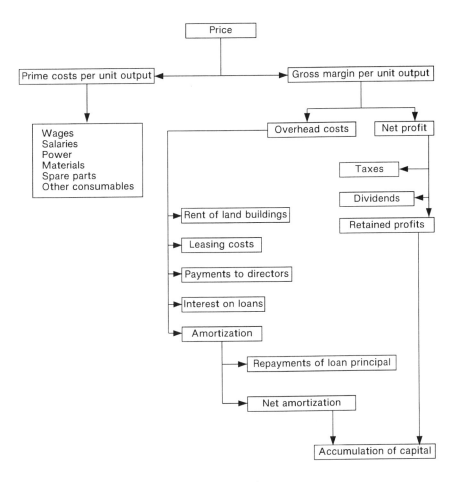

Figure 3.2   Price, prime costs and the gross margin.

3.4 Economies of scale

Neutze has suggested that three key characteristics of infrastructural investment are durability, specialization and immobility (Neutze 1994, 1997). These accord with the productive phases of the hydrosocial cycle. Another infrastructural attribute is often argued to be economies of scale. This claim requires careful examination in the case of fresh- and wastewater services supply.

Economies of scale can be said to exist when, over any given range of the average total cost function, cost per unit is lower at higher levels of output. In the case of the short run, we can confidently expect such economies to exist from the lowest output levels up to high levels of capacity utilization. However, as a plant approaches its maximum usable capacity, short-run diseconomies of scale may begin to set in. Figure 3.1 illustrated scale economies right up to usable capacity and, for most practical purposes, one can assume this to be true, in which case we have a **hyperbolic function**[2] for the short-run average total cost curve.

One now turns from short-run scale economies, where output change primarily derives from operational decisions, to long-run scale economies where capacity is raised through the investment process. The best approach to the long term is to consider it as an *ex ante* calculation, a view of future possibilities not yet realized, a planning approach.

With the long-term cost function, the starting point is a real catchment area, at a specific time, with all the infrastructural investments in fresh- and wastewater services inherited from the past in place. The interest here is in ATC at each alternative value of a set of output volumes derived from additions to capacity, from a minimal increment through to any upper limit one believes to be appropriate. Figure 3.3 represents such a function.

Note that the curve does not consist of a set of points representing how ATC changes as the scale of production expands over time. It is not an evolutionary function. What it represents, for each output level, is what ATC would be if, in a single leap, capacity were expanded up to that level. It is referred to as a long-term supply function not because the chart represents economies of scale with the passage of time but because the long term is the conventional timescale over which additions to capacity can take place, in contrast to the operational changes of the short term. To put it in Marshallian terms, the long-term supply function is an envelope constructed from the ATCmin of each of a series of short-run supply functions.

The hypothetical function in Figure 3.3 shows clear economies of scale up to 140 units of output per day. However, here, diseconomies of scale appear above 150 units per day.

This cost function is quadratic in nature, that is, well represented by an equation of the form:

---

2. In this case, depicted as a curve on a graph, the *y* value falls continuously as the *x* value increases, but by ever smaller amounts. Thus, the *y* value approaches a mathematical limit.

$$ATC = aL^2 + bL + c \tag{3.1}$$

where L is the output quantity. In the case of Figure 3.3, the equation is:

$$ATC = 0.004L^2 - 0.998L + 93.5 \tag{3.2}$$

How can one explain long-term economies of scale in fresh- and wastewater services provision? It is the measure of what economists widely refer to as the **indivisibility** of infrastructural provision. This neatly expresses the idea that, to secure even modest levels of output, major works are necessary – a sewer is a good example – and these are negligibly less costly than works which secure more substantial output levels. As a result, ATC falls markedly. The existence of significant scale economies at low to medium output levels has a most important social effect. It implies that, at least for small- and medium-size urban areas, freshwater supply and

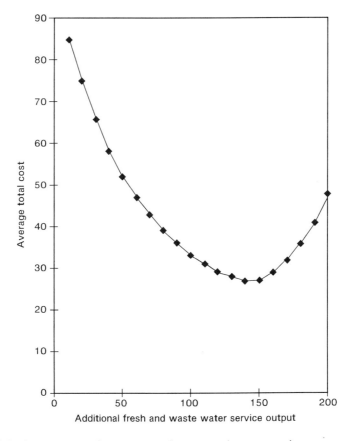

Figure 3.3 Long term supply: average total cost per unit output per day.

wastewater services constitute what is widely called a natural monopoly. The same can hold true in infrastructural provision for transport, energy supply and telecommunications.

Neutze stresses that unambiguous scale economies exist for the networks of land drainage, flood control and fresh- and wastewater services. The cost of pipes increases roughly in proportion to their length and diameter, "but their capacity increases in proportion to their diameter to the power of about 2.6 because the cross-sectional area increases as the square of the diameter and because friction between the water and the pipe decreases with size. . ." (Neutze 1994: 7)

But production efficiency is not a narrowly technological question. If we recognize that, in most industrial countries, fresh- and wastewater services supply at the catchment or regional scale is a monopoly, then economies of scale for the monopolistic firm may also derive from spreading administration, personnel, legal, monitoring, research, development, financial and accounting staff costs over higher output levels (Adams & Brock 1994: 101).

How, then, can one explain long-term diseconomies of scale at high levels of capacity? One possibility is that larger output levels require distribution networks for fresh water and collection networks for wastewater in areas where population distribution is less dense, raising piping and pumping costs more rapidly than output. A second reason of great importance in some countries, such as those in the Middle East, is that, as the abstraction of water increases in scale, it becomes more and more expensive to bring it to a population centre, as engineers have to tap supplies at a greater distance or in deeper aquifers. It may be, too, that greater output volume necessitates the use of water of lower quality, entailing higher treatment costs. One such case is where it becomes necessary to shift from fresh water to saline water abstraction.

In the discussion above, it was stressed that the long-term cost function of Figure 3.3 deals with additions to capacity "in a single leap", that evolutionary development is not in question here. However, when engineers are preparing plans many years ahead, it may be necessary to plot an evolutionary path. In the case of a ten-year water strategy, for example, it may make sense to consider two rounds of investment.

Here one is concerned not with a once-for-all heave to arrive at a given output, as with the long-term supply function, but with a first investment commitment followed by a second round of construction, for example, as an urban area's demand expands over time. Average total cost for the first of the two investment rounds is defined in Figure 3.3 in our example. But the ATC of the second round is not so defined, since the long-term cost function will have been shifted as a result of the prior introduction of the first-round infrastructures. Long-term supply costs are spatially and temporally specific.

Here is an example. Suppose that the stock of water in a reservoir for a town, when it is fully developed, will need to be 100 units. The single-round solution might simply be to build a 100 unit facility. A two-round solution might be to construct a 75 unit reservoir initially and then to add 25 units more in a second stage.

The single-throw approach may have the advantage of providing a more cost-effective solution when the town is fully developed but at the cost of under-utilized capacity during the first phase of the town's development.

But note, too, that the second-round addition of 25 units might be achieved either by a free-standing 25 unit reservoir or by deepening the 75 unit facility. With a double-throw, the first investment round changes the options available when the second round arrives. So, we can define the indivisibility of water infrastructures as what makes a single 20m dam cheaper than two dams each of 10m – a cross-section comparison. The **lumpiness** of water infrastructures is what makes a 10m dam built now impose a higher construction cost for an additional 10m capacity a year later – a time-series comparison. Neutze believes that "lumpiness" particularly characterizes the networks of fresh- and wastewater pipes. He argues that lumpiness occurs because land prices will have been pushed up by first-round infrastructural construction; because pipe installation is more expensive when it necessitates digging up an existing set of pipes; and because second-round investment is more disruptive than that on virgin sites (Neutze 1994: 4).

If the range of second-round options are part of a long-term infrastructural plan in designing the first-round facility, the 25 unit expansion is likely to be achieved more cost-effectively than without such planning. This is the basis of Neutze's argument that the durability, specialization, immobility, indivisibility and lumpiness of infrastructural provision provide a strong economic argument for urban land-use planning and the demand management of urban growth. To this one can add that, if second-round investments enjoy lower average total cost by using first-round structures, the natural monopoly of supply will be reinforced. To continue the previous example, this would hold true if deepening a 75 unit reservoir to increase its stock of water by one-third is cheaper than constructing a new free-standing 25 unit reservoir.

3.5 The cost of quality

Up to this point, output change has been considered strictly from the point of view of the cost increases accompanying expansion in the quantity of services provided. But output also changes as higher standards, incorporated particularly into fresh- and wastewater treatment, raise the quality of the product. The global shift in seeing water as an economic resource is attributable not only to the privatization of the water industry in different parts of the world but also to higher standard-setting, in response to environmentalist pressure and subsequent government action. These new standards make the hydrosocial cycle a far more costly activity in financial terms than it has been in the past, certainly in Europe.

Quality is measured in physical, chemical and biological terms and is assessed for surface water, groundwater, drinking water, wastewater and sewage sludge, especially when sewage sludge is used as a soil conditioner or fertilizer. Within the

European Union, it is the European Commission's periodic directives that have been the immediate driving force for quality innovation. In this context, the investment process in the EU's water industry incorporates technological change not so much as a form of interfirm rivalry (the standard textbook case) but as a response to the legislative powers of supranational government.

The relationship between cost and quality can be expressed in more than one way. For example, one can represent the planning function of Figure 3.3 not as a relationship between ATC and output per day, for freshwater quality of a given specification, but as one between ATC and a series of quality categories, each one more stringent than the last, for a given scale of output. Alternatively, one could plot a graph, for an existing plant of given scale and output quality, on how ATC would rise as a result of each of a set of options to raise quality. Again, this would be a relationship between a continuous variable – cost – and a categoric variable – quality.

3.6 Case study: sewerage costs in England and Wales

In England and Wales, at the end of the 1980s, there were just ten public sector regional water authorities supplying both water and sewerage services. They provided water to about three-quarters of the total of water customers in the two countries, as well as a sewerage service throughout their areas. As part of Prime Minister Thatcher's privatization programme, the 1989 Water Act transferred the **assets**, **liabilities** and functions of these authorities to ten private sector companies registered under the Companies Act 1985. Each of these was a subsidiary of a water holding company (Ofwat 1994a).

The Act radically changed the structure of the fresh- and wastewater industry, and provided for regulation by three bodies: the National Rivers Authority (NRA), the Drinking Water Inspectorate, and the Office of Water Services (Ofwat).

The NRA was charged with the protection and improvement of the water environment in England and Wales, and its responsibilities included pollution control, fisheries, conservation and recreation, and management of water resources, as well as a responsibility for a number of navigations.

The Drinking Water Inspectorate ensures that water companies comply with their statutory duty under the 1989 Act, later consolidated in the Water Industry Act 1991, to supply wholesome water. This is done by checking the results of the sampling and analysis that companies are required to carry out under the Water Supply Regulations 1989. Where standards have not been met, water companies have entered into legally binding undertakings to install additional treatment.

Ofwat is a non-ministerial government department that supports the Director General of Water Services, the economic regulator of the water and sewerage industry. The Director's duties are to ensure that the 10 new private companies created in 1989, as well as 21 water-only companies, can carry out and finance the functions specified for them in the 1991 Act. Subject to that, the Director must "protect customers, promote economy and efficiency and facilitate competition".

Each of the ten companies operates under a licence granted to it by central government, and Ofwat monitors them to ensure that they comply with their licence conditions. Ofwat compares their performance in order to encourage greater efficiency and value for money for customers, and also monitors the effectiveness of investment programmes to see that companies are delivering the improvements promised when charge limits on water and sewerage services were set. Companies must keep the average percentage rise in the fixed charge to customers within a limit set as the rate of change in the retail price index (RPI) plus a factor called K. Initial limits were set in 1989 and updated in 1995 (Ofwat 1994a: 5–6).

With respect to the sewerage function, the ten private sector companies regulate, by means of consents and agreements, the discharge of effluent from trade or industrial premises into the sewerage system. They have powers to charge for disposing of such effluents, and to set quality standards so as to protect their own assets and to ensure that the discharges from their sewage treatment works meet the standards prescribed by the NRA. Discharges of trade effluent containing the most dangerous substances (the "Red List") are referred to central government to determine appropriate consent conditions. Her Majesty's Inspectorate of Pollution (now part of the Environment Agency) sets those conditions.

To carry out its monitoring work and its investigations of comparative efficiency, Ofwat has research staff located in its Charges Control Division. Their studies of sewage treatment works have become particularly important since the EU adopted in May 1991 its Urban Wastewater Treatment Directive. This lays down uniform emission standards for sewage treatment from all sewage treatment works serving populations of 2000 or more. Ofwat has estimated that over ten years the Directive could add an extra £10200 million to the ten companies' capital investment programmes and £2400 million to operating costs (Ofwat 1993, 1994b). (In October 1995 the rate of exchange of the pound sterling to the dollar was 1:1.57.)

The 1991 Directive specifies the dates by which works of a given size must comply with the requirements, with all work to be completed by 2005. Capital expenditure is likely to be needed at 1125 sewage treatment works in England and Wales, and at 265 crude sewage **outfalls**.

As Ofwat states (1994b: 1):

The purpose of the Directive is to protect the environment from adverse effects of sewage discharges and to ensure that all significant discharges are treated before being released to inland fresh water, estuaries or coastal waters. Sewage will normally receive **secondary biological treatment**, but in some coastal areas, where there is a high natural dispersion of the discharge, **primary treatment** (involving settlement of solids) is considered acceptable. Discharges into sensitive areas will require more stringent treatment. Among other things, the Directive requires an end to the dumping of sewage sludge at sea. The Directive also covers discharges from industry into rivers, estuaries or the sea.

Beginning in 1992, Ofwat has published each year a report that presents the costs of providing water and sewerage services. The costs of sewage collection, treatment and disposal are expressed in pence for each cubic metre of sewage collected. The volume of sewage collected is considered to be the most appropriate output measure of the sewerage service as a whole, as it is relevant to costs of both sewerage operations and sewage treatment. However, it is not considered to be a robust variable. Sewage collected is difficult to estimate; some sewerage systems are used for surface water drainage; and sewage volumes are not the only driver of sewage treatment costs.

In any event, costs are classified under four heads (Ofwat 1994c: 8–11): first, the cost to customers is the average revenue received from sewerage customers; secondly, the cost of operations covers such items as staff wages and salaries, and spending on power, materials and hired and contracted services; thirdly, the cost of capital maintenance principally comprises **depreciation** and an infrastructure renewals charge, the latter representing the cost of maintaining the operating capability of infrastructural (mainly underground) assets; fourthly, the return on capital is operating profit before interest.

In Ofwat's reports these costs are so defined that cost to customers is exactly equal to the sum of cost of operations, cost of capital maintenance and return on capital. We cannot speak of a price to customers, as the water and sewerage industry derives its income primarily from fixed charges.

Table 3.1 shows the cost information for Thames Water plc, one of the ten private companies already mentioned. The cost to customers in 1993–4 was £0.54 per m^3, in contrast with the weighted average for England and Wales of £0.75 per m^3. This average had grown by 7 per cent in real terms over the previous year's datum, reflecting increased expenditure to fund improvements in effluent quality to meet standards set in the 1974 Control of Pollution Act and in the EU's Bathing Waters Directive (Ofwat 1994c: 13).

As part of the assessment of companies' comparative operating efficiency, Ofwat (with Professor Mark Stewart of Warwick University) has developed five sewerage service models. These are (ibid.: 38–43):

two detailed econometric models of sewerage and large sewage treatment works based on data below the company level and three less sophisticated analyses of small sewage treatment works, sludge treatment and disposal, and other business activities. These models were combined to derive an overall measure of each company's comparative efficiency.

The large sewage treatment works model used 1992–3 company data on 333 individual plants with an average daily load exceeding 1500 kg **BOD5**, equivalent to a population of about 25 000. Data from the ten plcs were amended, for example to exclude works with untypical treatment and to remove cost duplication. No direct measure of load was available, so an estimated total was derived from the size of the resident population in the area (60 g pppd), the size of the non-resident,

Table 3.1 Sewage collected unit costs 1993–4 for Thames Water plc (pence per m³).

Cost to customers	Cost of operations	Capital maintenance	Return on capital
54	23	12	19

Source: Ofwat (1994c: fig. 2).

holiday, population in the peak month (20 g pppd) and the trade effluent and tanker loads treated. In the case of Thames Water, in 1993–4 the total volume of sewage collected was 2550 Ml per day. The unmeasured household volume in litres pppd is estimated to be 147, whereas the unmeasured non-household volume is estimated at 1602 litres per property per day.

The econometric model for large sewage treatment works is of a logarithmic form, which is particularly appropriate where economies of scale exist, and it finally included ten independent variables that were shown to be statistically significant in explaining costs. The R^2, that is, the proportion of the variation in cost explained by the model, was a very satisfactory 0.81, given that the load data could not have been completely accurate. In its final form, the equation was (Ofwat 1994c: 40):

Large works functional expenditure less terminal pumping costs =
0.401
$\times (\text{load})^{0.71}$
$\times e^{(-0.16 \times \text{ratio of trade effluent to total load})}$
$\times e^{(0.0068 \times \text{distance to next works})}$
$\times 2.39$ if a secondary activated sludge works (with or without tertiary treatment)
$\times 1.34$ if a secondary biological works (with or without tertiary treatment)
$\times 1.21$ if a secondary biological works with tertiary treatment
$\times 1.40$ if a secondary biological works with a BOD consent of 20 mg per litre or less
$\times 1.30$ if an activated sludge works with tertiary treatment and an ammonia consent of 5 mg per litre or less
$\times 1.58$ if the works functional expenditure includes costs of treating only its own sludge
$\times 1.75$ if the works functional expenditure includes cost of treating sludge from its own and other works

The beauty of the model, from the point of view of this chapter's focus, is that it identifies empirically the long-term scale economies of indivisibility, albeit using cross-sectional data rather than the estimates of an *ex ante* planning function. Thus, a load difference of 10 per cent brings with it only a 7 per cent difference in additional functional expenditure.

Moreover, the model also measures the cost of quality. Thus, the fifth independent variable in the equation above shows that a works with secondary biological treatment would be expected to have costs 34 per cent higher than an equivalent primary works. The sixth variable shows that, if the works also had tertiary treatment, costs would be 21 per cent higher than the secondary biological works, that is, 62 per cent higher than the equivalent primary works. "Activated sludge works are shown to be more expensive than biological works, and tight consents, as expected, also impact on costs." (Ofwat 1994c: 41).

Finally, the model was used for each of the ten water and sewerage companies to predict operating expenditure for comparison with their actual expenditure. The difference between the two is expressed as a percentage of predicted cost. In the case of Thames Water plc, large sewage treatment works' actual costs were 19 per cent more than expected, suggesting significant inefficiency in Thames operations compared with other plcs (ibid.: table 9). It is worth noting that Ofwat knows of no convincing evidence that relatively high operating expenditure can be simply explained by relatively low capital expenditure. So, such a trade-off could not legitimately be used to explain away high operating costs.

3.7 Case study: tradable abstraction rights in Victoria

Chapter 2 suggested that the first stage of the hydrosocial cycle is abstraction. In this case study for Chapter 3, the issue of tradable abstraction rights in general is reviewed, and material available from Victoria in Australia in particular.[3]

In any specific catchment, some users of water may value abstraction rights more than others, for example because they employ water more productively. In this situation, trading in such rights can be financially attractive, both to the existing owner of such rights and the party that wishes to see them transferred. To understand trading in abstraction rights, then, we need to come to grips with the relevant law, with differential valuations of water, and with the legal, hydrological and engineering means for implementing transfers. In this field, valuable work has been done by Robert Hearne and William Easter, in Chile, although less than 1 per cent of all abstractions operate through such water markets there (Hearne & Easter 1995).

In Australia, the practice is far more extensively developed, at least in terms of total area, and there the main objective of tradable abstraction rights has been the more efficient use of scarce irrigation water. The Australian water economy is in its mature phase, where the long-term average total cost function is rising sharply, where there is intense competition for existing supplies, and where the hydrosocial infrastructure requires costly rehabilitation (Randall 1981; Pigram et al. 1992: 3).

Victoria is one of the constituent states of federal Australia and lies in its south-

3. The section is based on the work of John Pigram and his co-authors, and of Henning Björnlund & Jennifer McKay. Pigram is the Director of the Centre for Water Policy Research at the University of New England in New South Wales, whereas Björnlund & McKay are based in the Faculty of Business and Management of the University of South Australia.

east corner with South Australia to the west, New South Wales to the north and east, and the sea to the south (Fig. 3.4). The dominant physiographic feature is the Great Dividing Range, running east–west in the eastern two-thirds of the state. This creates a rainshadow in northern Victoria, where average annual rainfall is 450 mm in the Goulburn–Murray Irrigation Area (GMIA). This is the largest irrigation area in Australia, covering some 820 000 hectares, and it represents the bulk of water trading in the state. Irrigation is sourced from storages on regulated surface-water flows. Meat and dairy products are the principal agricultural output; higher-valued

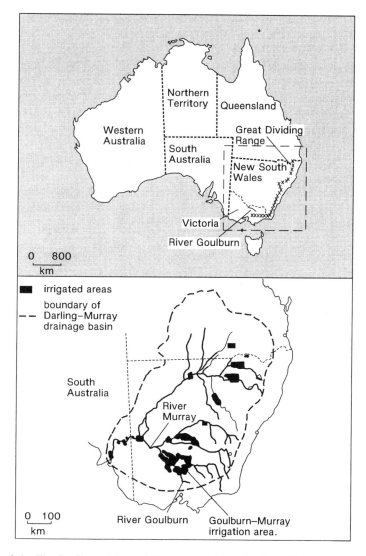

Figure 3.4   The Goulburn–Murray irrigation area, Victoria, Australia.

crops such as horticultural products are constrained by the red-brown soil type. Surface water resources have been extensively developed in the past and any further supply growth would require inter-basin transfers. The GMIA is principally designed to provide security of supply during a prolonged series of drought years. The main water storages are operated on a carry-over basis, where water is accumulated in years of high river flows for use in drier years (Pigram et al. 1992: 91–9; Björnlund & McKay 1995).

Australia's modern history of property rights in water begins in the 1880s, when Victoria introduced new water laws based on the recommendations of a Royal Commission chaired by Alfred Deakin. Deakin proposed that water allocations should be tied to the land, that rights to water should be vested in the British Crown, and that allocations to landholders should be the responsibility of the state governments. Riparian owners retained limited common-law rights for domestic use, stock watering, gardens and a maximum of 2 hectares of irrigated land for fodder crops. Over the next 15 years, similar legislation was introduced in the other federal states. Using the new legislation, the states ventured into large irrigation projects, providing water to farmers far below average total cost. Areas were established in which government constructed massive irrigation and drainage infrastructures, such as the GMIA, the Murrumbidgee Irrigation Area in New South Wales, and Riverland in South Australia (Björnlund 1995).

The subsidy of water led to over-use and this was consolidated by the historical over-allocation of water to irrigated farms. As argued in the work of Pigram et al. (1992: 6–7):

(This) has its roots in the social objectives of past governments for the development and use of water resources. The overriding objectives were an equitable distribution of water among farms and the promotion of regional development and closer settlement of inland areas. All farmers were considered to have an equal right to the available water, irrespective of how much was needed to irrigate the proposed crops to be grown on the farms, or of any consideration of how efficiently the water would be used. As water rights became capitalized into land values, individuals had the economic incentive to retain their entire water rights through demonstrating a history of use. In many cases, this actually translated into a history of over-use.

By 1990, the arrangements in the GMIA were that irrigators paid for a water right based on both the amount of land they held suitable for gravity-fed irrigation and the irrigation district in which they were located. An annual charge applied to the water right, invariant with the volume of use. Subject to availability, "sales water" was also available, at a volumetric charge based on the water right charge. These water rights in their specific form were assigned by a licensing system for a fixed period of 15 years, with an expectation of reissue. The volume licensed could be varied by the authorities, usually in times of shortage. However, before 1987–8, no arrangements existed for transferring farmers' abstraction rights.

An Australian interest in tradable abstraction rights, also known as tradable water rights or transferable water entitlements, begins to surface in the mid-1970s. For some, the introduction of tradable abstraction rights seemed to offer a new flexibility to existing arrangements, with clear economic and environmental advantages (Pigram et al. 1992: 5–9).

First, we have already seen that the maturing of the Australian water economy in recent decades is associated with high average total cost for the long-term supply curve. So, tradable abstraction rights offered a new direction which, by redistributing water supply, would reduce the pressure for aggregate supply expansion. This redistributive effect would not be merely within the farming sector. There was also the prospect of reducing agricultural over-use to open up supplies for a range of urban and industrial uses.

Secondly, supply redistribution within agriculture was, of course, a major objective. There was a widespread belief by the mid-1980s that tradable abstraction rights would switch water from lower to higher water-productivity uses in the farming sector. In Australia, at that time, the agricultural sector accounted for 80 per cent of total water use. In Victoria, each year, up to a third of irrigators were using less than their full water right allotment. Specific switches into river red gum watering, salinity dilution and dairy farming were forecast. (The variability of agricultural **water productivity** will be addressed again in this book in case studies of Jordan (Ch. 4) and California (Ch. 7)).

Thirdly, environmental benefits were expected from transferability. The new policy promised to reduce the scale of infrastructure construction for interbasin transfers, with all their negative externalities. Development proposals were meeting increased resistance from environmental groups (Björnlund 1995). Moreover, reductions in agricultural over-use can have a positive environmental effect through a reduction in waterlogging and salinity. Excessive water use also results in increased **runoff** and return flow to the rivers, containing salt, chemical fertilizers and biocides (Pigram 1986).

The redistribution of abstraction rights could have been sought by the administrative processes of the licensing system. But this would have been met by political resistance from the farms on which volumes were to be reduced and land values consequentially cut. Tradable abstraction rights offered these farms a payoff. As Pigram et al. put it (1992: 9):

For landholders wishing to move out of irrigation, but to remain in dryland agriculture, transferability allows water entitlements to be sold separately from the land. Previously, the options for such irrigators were either to cancel their water rights licence (or not renew it) and get nothing for the right, or sell the entire irrigation holding and buy a dryland property elsewhere. Transferability can permit a more flexible retirement plan for an irrigator, or facilitate a long-term change in enterprise or financial structure of the farm business.

The fundamental requirement of a workable and efficient market in abstraction rights is a clear specification of property rights in water, such that: first, rights in land and rights in water must be separable; secondly, the volume of water that an individual or institution has available for transfer must be clearly stated, as well as any special conditions on its use – such volumes are likely to be conditional on effective rainfall, or surface flows or groundwater stocks; thirdly, the right to transfer such water at a privately negotiated price must exist; fourthly, the period over which such a transfer is deemed to be effective, temporary or permanent, must be known; fifthly, the power of government to restrict or terminate abstraction rights at a future date must be known.

In Victoria, tradable abstraction rights were cautiously introduced in 1987–8 with a temporary scheme, before permanent transfers were considered. In the Goulburn–Murray Irrigation Area, the arrangements were as follows: the transfer period would be for one year, only between irrigators, and only within the same supply system; there was no volumetric limitation, but stock and domestic allocations had to be retained; the relevant state agency assessed possible third-party effects and could refuse the transfer if these were significant; the arrangements should not significantly affect delivery and drainage channel capacity or salinity; the agency's fee was Australian $70; and the price was determined between buyer and seller (Pigram et al. 1992: 24).

A new Water Act in 1989 permitted permanent transfers of abstraction rights between farmers, with effect from 1991–2. Transfers out of agriculture were still proscribed. However, despite these references to "permanence" in trading, the state assignation of water rights through the licensing system, described above, meant that a purchaser of abstraction rights could not be assured in law that such water rights would continue indefinitely. The state government retained long-term flexibility in its management powers.

Tradable abstraction rights impose certain transaction costs, that is, economic costs necessary to transfer a right in property from one party to another. As has been said, the state agency demands a $70 fee. Farmers themselves also have to incur some legal costs. Incremental supply costs may also exist, if distribution and drainage structures cannot handle the increased water volumes. However, in Australia, water agencies have taken the easy administrative option of simply refusing transfers where existing capacities would be exceeded.

Next, the outcomes of Victorian tradable abstraction rights are considered, almost ten years after they were first introduced. In terms of scale, this can be measured by the volume of abstraction transfers as a percentage of all abstractions. With respect to benefits, we are looking for *ex post* evaluation of the claimed advantages of tradable abstraction rights already referred to above, that is, a saving of infrastructure expenditures on new supplies made redundant as a result of the redistribution of abstractions within and between user categories, as well as the avoidance of such infrastructures' negative externalities; an increase in agricultural productivity as a result of abstraction redistributions within farming; and a fall in waterlogging, salinity and pollution from runoff because of the reduction in agricultural over-use. (Note that the economic and environmental evaluation of water projects is considered in Chs 5, 6 and 8 of this book.)

Pigram et al. showed that, with respect to the volumetric scale of traded abstractions in Victoria (most of it in the GMIA), in the two years 1987/8 and 1988/9, the annual average was 25 million m³, less than 2 per cent of total abstractions. They argued that the true magnitude of benefits would only be evinced in severe drought conditions. This seems rather dramatically to undermine the general case for tradable abstraction rights. Price per megalitre ranged from Australian \$8–20 (1992: 26–7, 103). With respect to economic benefits, Pigram was reporting too early for much *ex post* evaluation to be possible. The potential benefits were assessed in terms of additions to net farm income and, as expected, these were positive. However, again, the comment is made that "Despite widespread endorsement of the concept in Victoria, transfer activity has been sporadic in that state." (ibid.: 43 and ch. 6).

Björnlund & McKay wrote in the mid-1990s, when enough time had passed for the new arrangements to bed in and for permanent transfers to have taken hold. But in Victoria by 1990/91 trade was still "insignificant compared to the total volume" (Björnlund 1995). In a mail survey of all 299 permanent water transfers within the GMIA up to 1994, with a 63 per cent response rate from buyers, purchases made up between zero and 2.3 per cent of total allocations when the responses are grouped by district (Björnlund & McKay 1995: table 2). Stringer (1995) suggested that in Victoria as a whole in 1991–4, the average percentage of allocated water traded in each of those years was 0.33 per cent. In the Murray region as a whole, in 1993–4, the volume of transfers in millions of cubic metres, was 17 for temporary transfers and 8 for permanent transfers. For 1994–5, these figures changed dramatically: temporary transfers soared to 265 million m³, while permanent transfers fell to 2 million m³ (Pigram: pers. comm.).

Björnlund & McKay have not published any *ex post* evaluation of the economic benefits of Victorian tradable abstraction rights, nor does there appear to be any published environmental impact analysis of trading there. What the GMIA seller/buyer survey does show is that sellers are releasing sleeper water (i.e. unused allocations) and that farmers in financial distress are also selling. In a smaller number of cases, farmers either wished to cut their irrigation agriculture, or to farm without irrigation, or to retire. In respect of purchasers, the single most important reason for buying water, applicable to 65 per cent of respondents, was "wanted to secure existing crops against future drought" (Björnlund & McKay 1995). This is consistent with the remarks already quoted above of Pigram and his co-authors.

In summary, particularly in light of the 1993–5 data for the Murray region as a whole, tradable abstraction rights in Victoria seem to have created a space for single-season switching of abstraction, most active in years of low rainfall.

3.8 Final remarks

The principal objective of Chapter 3 has been to consider the supply costs of fresh- and wastewater services. Water's real costs are pre-eminently the domain of the engineer, the chemist and the hydrologist, and range from abstracted water itself, through land sites, infrastructures such as reservoirs and works' roads, treatment plant, the distribution and collection networks, electric power, materials and the human labour required to design, build, operate and manage the whole system.

On the basis of market prices, the economist represents these real costs in money terms, so that they can be treated arithmetically. An important technique of cost classification is their separation into capital and current expenditures. These two categories are vital in considering how production is to be financed, and in the evaluation of investment options.

A related and complementary cost classification is into prime, overhead and total costs. These can be used when we examine how costs per unit of output (the cost function) vary with the scale of output in the short term, during which time new investment for capacity expansion is ruled out.

Short-term cost analysis is particularly useful in assessing current-period profitability. The excess of product price over average prime cost defines the firm's gross margin. The excess of price over average total cost defines net profitability and is the source of taxes paid to government, shareholders' dividends and net investment. Figure 3.2 shows the various pay-offs financed by the price received per unit output. The short term is marked by strong economies of scale, that is, lower average total cost as production expands towards the usable capacity limit, and the shape of the short-term function is hypothesized to be hyperbolic.

The long-term cost function is constructed as a planning concept. It is defined as the average total cost per unit output for future values of additional fresh- and wastewater services supply capacity. Here ATC is calculated for each discrete incremental output level on the assumption that capacity is achieved in "a single leap" from the base-period situation. The hypothesis is that this relationship between ATC and additional capacity is well represented as a **quadratic function**. At low and medium scales, ATC falls because the hydrosocial cycle's infrastructures are characterized by indivisibility, in both engineering and staffing senses. This is the basis of the natural monopoly of water and sanitation services. However, at high output levels, ATC rises as freshwater supply sources become more costly to access and as their quality diminishes. The indivisibilities of the long-term cost function, and the lumpiness of evolutionary development in a catchment, call for urban and regional planning of the water and sanitation sector.

The analysis concludes by pointing out that output shifts can be qualitative as well as quantitative and that, in the European Union, investment since the mid-1970s has primarily been quality-driven, as national governments and the European Commission have responded to environmentalist pressures. The first case study shows both scale economies in sewerage headworks, as well as the costs of quality in sewage treatment. The second case study, of tradable abstraction rights in Victoria, Australia, shows that there (as well as in Chile) the scale of trading is extremely modest, that such rights are predominantly temporary, and that they are of greatest importance during dry seasons.

CHAPTER FOUR

Effective demand and the price of water

4.1 Introduction

Chapters 2 and 3 concerned themselves exclusively with supply. The time has come to focus on the demand-side of hydroeconomics. The consumption of fresh water is most conveniently analyzed from the point of view of the three large battalions: households, farmers and industry. By "industry" is meant, here, all producers of goods and services in the economy other than within the family and within agriculture. So, industry includes, for example, an enterprise manufacturing nitrogen fertilizer, a company generating electricity, the head office of a banking group, a publicly owned rail network, a local hospital, a newsagent . . . and so on and so forth.

Personal and household needs for water are discussed at length in the box on the next page. In agriculture and industry the issue is analytically simpler. In both cases, water is a raw material required for the production process. In farming, water is a biological necessity for the growth of plants and the raising of livestock. In industry, water fulfils a multiplicity of functions, including cleaning, cooling and power generation.

4.2 The concept of effective demand

At this juncture, it is appropriate to highlight the distinctions between three concepts of water use. The first of these is the *need* for water, already referred to above. Debates on need concern why human society finds useful or derives some form of satisfaction from a commodity. The analysis here is primarily textual rather than quantitative. Economists have no unique expertise in need analysis: it is also a field for the philosopher, the sociologist, the psychologist, the market researcher and – in the case of water – for the biologist, the agronomist and the industrial consultant (Gough 1994).

Our need for water

Households' need for water is complex. At the most basic level, there is a bio-logical drive, a physiological necessity. Some 60 to 70 per cent of the body weight of the average adult human being is constituted by water molecules. The author of this book – to take a ready example – is constituted by at least 36 litres of water, using the approximate equivalence of one litre of water to one kilo-gram. Specific functions of water in the body include facilitating the chewing, swallowing and digestion of food; the transport and disposal of body wastes; and the production of blood. Generically, and most importantly, water is the basic solvent necessary for all the body's functions, both at the biochemical and the cellular level.

A 70 kg man on an average diet has a daily water intake of about 2 litres, of which some 56 per cent is derived from liquids, 22 per cent from solid food and 22 per cent from the oxidation of foodstuffs. Liquid intake may be in the form of drinking water itself, or more sophisticated products such as tea, coffee, beer, wine, milk, soft drinks and so forth.

In the normal adult organism, not undergoing changes in weight, the quantity of water supplied daily is balanced by that eliminated, a state of equilibrium being maintained. The daily water output of an adult is typically composed of 53 per cent as urine, 42 per cent as evaporation from the skin and lungs and 5 per cent in faeces.

This accounts for the biological need of the individual for water. But as members of a family we also need water for a multiplicity of other purposes, such as cooking, washing ourselves, disposing of body wastes through the flush-toilet, cleaning, and watering the garden. Food preparation, personal cleanliness and the disposal of urine and faeces all have a biological dimension. For example, an important cause of diarrhoea is preparing and eating food with unwashed hands. But the uses of water within the household also have a cultural basis, as the washing of hair before a date, the running of a dishwasher, and cleaning the car all demonstrate. The *forms* in which we ingest water is also a cultural-economic product, in part because of the importance of drink manufacture within the economy.

For these reasons, the supply of water to households can be seen as a form of *reproductive production*, alongside the securing of food, the construction of accommodation, the making of clothing, and the supply of health care and edu-cation services (Merrett with Gray 1982: 72). This is because these six sectoral processes of economic *production* provide goods and services that are directly necessary for the *reproduction* of the human species. Each is simultaneously both a form of human labour and is a direct and necessary condition for the reproduction of the labour force. These six productive sectors are the oldest forms of human economic activity.

The second water-use concept is that of *consumption*. This refers to the quantity of water used by a single consumer, or consumers in aggregate, in any given time-period. Consumption may also specify the quality of the water used. It is often represented in graphs where, for example, cubic metres per unit period of time appear on the vertical axis, and the horizontal axis measures the passage of time. Engineers are often apt to refer to consumption as "demand". *In this book, the terms "consumption" and "use" will be used interchangeably.*

The pattern of consumption varies markedly across the world. For example, Kinnersley writes (1994: 181):

In a large group of low-income countries, the domestic sector is estimated to take only 4 per cent of total water withdrawals and the industrial sector 5 per cent, while agriculture takes 91 per cent. The equivalent figures for high-income countries are 14:47:39.

The composition of intrasectoral consumption can also be surprising. In the UK domestic sector, for example, average daily consumption is 140 litres per person, divided as follows: toilet flushing, 32 per cent; baths and showers, 17 per cent; clothes washing, 12 per cent; other internal, 35 per cent; external, 3 per cent (ibid.: 94).

The third water-use concept is specifically that of the economist and is *effective demand*. Effective demand is the relationship, at a given time and within a defined market, between price per unit of a product or service and the quantity in each time period that consumers are estimated to be willing to purchase at each price. Effective demand is conventionally represented graphically, with price on the vertical axis and quantity on the horizontal axis, showing the difference in quantity purchased at each price.

The cultural and economic context of any effective demand function is referred to as the conditions of demand. The first of these conditions is the tastes and habits of consumers, that is, the nature of need for the commodity. The second is the price, quality and availability of commodities that consumers consider to be substitutes for the product. These two together should account for the consumers' *willingness* to purchase. The third condition is the incomes, assets and access to credit of consumers, which account for the *ability* to purchase.

Where the conditions of demand are stable, the graphical function can be used to represent not only price and quality *differences* within a given time-period, as above, but also price and quality *changes* in successive time-periods. This is legitimate only if expectations about the future are stable. Almost invariably, under these conditions, higher prices are associated with lower quantities and so the effective demand function slopes downwards from left to right, as is shown in Figure 4.1.

A matter of great interest in the economic analysis of effective demand is how responsive consumers are to product price differences. With any given difference (or change) in price, is the response large or small in terms of quantity purchased? One measure of this concept of responsiveness is called the price elasticity of demand and is equal to the proportionate difference in quantity purchased divided by the proportionate difference in price paid.

$$e = -\frac{(\Delta Q)/Q}{(\Delta P)/P} \qquad (4.1)$$

where e is the price elasticity of demand, Q is the quantity purchased, P is the purchase price, and ΔQ and ΔP refer to the quantity and price differences. Since the demand curve slopes downwards, elasticity will always be a negative number. By convention, economists sometimes omit the minus sign, but we do not do that here.

Where the effective demand function is mathematically continuous, e could be measured for each and every point. In practice, economists estimating the function construct a segmented "curve" as illustrated in Figure 4.1. The data there are drawn from the hypothetical case of Table 4.1. Because this function is discontinuous, we are measuring the elasticity between separate points on the curve. So, it is vital to

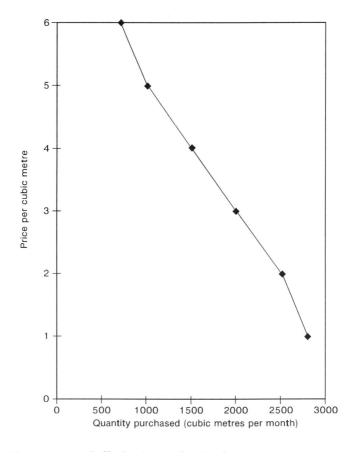

Figure 4.1 A segmented effective demand function for water.

make explicit whether the elasticity calculation of Equation 4.1 is based on the higher Q and P values between the two points, or on the lower values, since these two options give different values. Throughout this book, the higher values are used.

Table 4.1 The price elasticity of effective demand for water.

Price per m³	Quantity purchased (m³ per month)	Value of sales (col. 1 × col. 2)	Elasticity
6	700	4200	
			−1.80
5	1000	5000	
			−1.67
4	1500	6000	
			−1.00
3	2000	6000	
			−0.60
2	2500	5000	
			−0.21
1	2800	2800	

If we examine the fourth column of Table 4.1, it can be seen that the price elasticity of demand falls into three sets of values:

- For some price differences, such as between 2 and 1, the proportionate difference in quantity is *smaller* than the proportionate difference in price, and so the values of e lie between −1 and 0.
- Between prices 4 and 3 per m³ the proportionate difference in quantity is *equal* to the proportionate difference in price, and so the value of e is −1.
- For other price differences, such as between 6 and 5, the proportionate difference in quantity is *greater* than the proportionate difference in price, and so the values of e lie between −1 and minus infinity.

There is no implication here that low prices are *always* associated with low elasticities and high with high. Note, too, that with reference to e, "high" and "low" refer to the absolute value of demand elasticity. For example, −0.5 is a low elasticity and −3 is a high elasticity.

These relationships can be of great importance in shaping enterprise behaviour:

- Where e lies between 0 and −1, we speak of inelastic demand, and here a lower price brings a greater quantity purchased but with a lower value of sales. The 2:1 price difference in Table 4.1 shows this.
- Where $e = −1$, a lower price brings a greater quantity purchased, but with no change in the value of sales, as in Table 4.1's 4:3 price difference.
- Where e lies between −1 and minus infinity, a lower price brings a greater quantity purchased and with a higher value of sales, as in the 6:5 price difference.

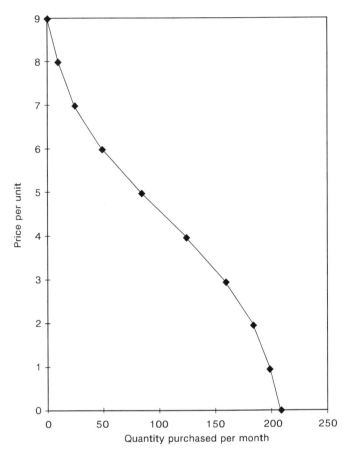

Figure 4.2   A cubic demand curve for water.

A general hypothesis I shall adopt about the shape of the effective demand curve for water is that it takes the form of a **cubic function**, where:

$$P = aQ^3 + bQ^2 + cQ + d \tag{4.2}$$

This is illustrated in Figure 4.2. At low quantities purchased, a higher price brings little reduction in the absolute quantity purchased, because of the intensity of need for the product. At middle-of-the-range quantities, a price difference brings a clear shift in the quantity purchased. At high quantities, a lower price eventually brings no increase in the quantity purchased because the consumer is satiated, not to say saturated, with water. The equation of the data recorded in Figure 4.2 is:

$$P = -0.202 \times 10^{-5} Q^3 + 0.001 Q^2 - 0.086 Q + 8.897 \tag{4.3}$$

4.3 Price determination in a free market

Let us now discuss how the price of water is determined, drawing on the supply analysis of Chapter 3, and Chapter 4's review of effective demand. Throughout, it is assumed that the supply of fresh- and wastewater services is a monopoly. The monopoly may exist for a specific town or city, or for an entire catchment, or for a region composed of several catchments. Initially, the analysis is limited by the assumptions that the water services company operates in a free market economy, where no state intervention is manifest, and that the focus of interest is the price of water supplied to households, industry and agriculture, not the supply price of wastewater services.

Since the first publication of Alfred Marshall's *Principles of economics* in 1890, it has been commonplace for neoclassical economists to theorize about price determination in a free market by using "Marshall's scissors". This is a diagram with a cross – X – at its heart, where a downward-sloping demand curve confronts an upward-sloping supply curve, and an equilibrium price is located at their intersection. The discussion by Robinson & Eatwell of scissors diagrams is valuable in this context (Robinson & Eatwell 1974: 161–74).

The analysis here will begin with just that traditional scissors approach, modified by the quadratic character attributed to supply (see Eq. 3.1) and the cubic character attributed to effective demand (see Eq. 4.2) The resulting diagram is Figure 4.3. In this case demand is in a specific year and for a defined catchment; it refers to the aggregate consumption of households, industry and agriculture; and it exists in a situation where the conditions of demand are stable. Quantity is measured in m³ per year purchased of water of a known quality. Price is in a given currency (e.g. US dollars per m³ consumed) and is assumed to be set at the beginning of each

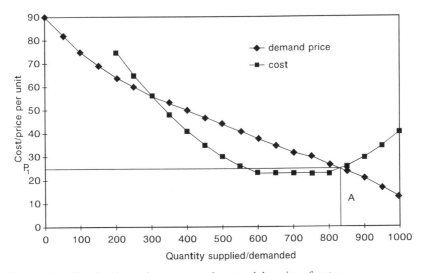

Figure 4.3   Effective demand, average total cost and the price of water.

financial year. The volumetric consumption of each customer is known because metering is universal. Price variations between customers are assumed not to exist, since that is not the focus of interest at this point. In respect of supply, the long-term function, like that in Figure 3.3, is used. Superimposing the supply curve on the demand curve provides an intersection at point A of Figure 4.3, which might be considered to produce the long-term equilibrium price P_1.

Unfortunately, this procedure is nonsensical. In the first place, the demand curve is for all water consumed, but the supply curve is only for additional capacity. Secondly, the demand curve is estimated for the financial year just beginning, but the supply curve is a planning function for an unspecified future year. Thirdly, a catchment monopoly is a purposeful institution which uses its own routines to set price. In that sense it enjoys free will. There can be no question of either a supernatural process by which price is set so as to solve the simultaneous equations of supply and demand; nor is it empirically the case that the monopoly's senior management actually uses that technique themselves.

The alternative theory adopted here is that senior management use an average cost-based pricing approach. Lee has suggested that this has three variations: target rate of return pricing, full cost pricing and mark-up pricing (Lee: 1994b). The last of these is used here and is illustrated in Figure 4.4. Key assumptions are that a single price for the three consuming sectors is set; that we are dealing with a catchment monopoly; that price is set for one year at a time, at the beginning of the financial year; and that metering is universal.

The pricing routine is as follows. The short-term average prime cost and average total cost functions are estimated for capacity utilization between, let us say, 40 and 100 per cent of usable capacity. These estimates are based on last year's experience plus any forecast changes for the coming year in items such as wages,

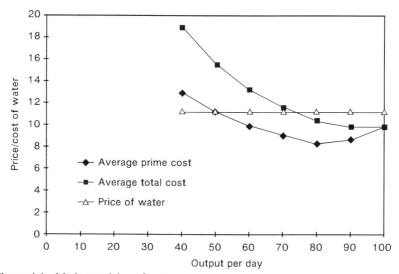

Figure 4.4 Mark-up pricing of water.

the price of power, loan rates of interest, and increases in payments to the board of directors. A normal rate of capacity utilization is assumed, once again based on past experience. Price is then set as a mark-up on the average prime cost at that normal level of capacity utilization. In Figure 4.4, the example assumes the normal level of capacity utilization is 80 per cent; that at this level average prime cost and average total cost are respectively 8.4 and 10.5 currency units per m^3; and that a mark-up of 35 per cent on average prime cost is set, giving a price of 11.34. That price, in this case, represents a level 8 per cent greater than average total cost.

The catchment monopoly may have the power to set what price it chooses, but it certainly does not have the power simultaneously to determine total volumetric consumption. If, for example, consumption is only 70 per cent of usable capacity, price will be 0.36 currency units less than estimated average total cost in Figure 4.4. If utilization is 90 per cent, price will exceed estimated average total cost by 1.34 units, and total net profits available for the taxman, for distribution to shareholders and for retention or new investment will equal 1.34 multiplied by the scale of output (see Fig. 3.1). During the financial year the true level of capacity utilization at the going price and of average prime cost and average total cost will strongly influence the following year's cost estimations and the pricing decision.

Now one can address the question of the monopoly's setting a separate price for each group of users – industry, households and farmers. In the water service company of the free-market **ideal type** considered here, that seems very likely. If the firm's accounting department demonstrates clear cost of production differentials for each consuming group, higher cost groups will be charged higher prices. A second source of price discrimination is the estimated elasticity of demand of each separate user group. A group with a low elasticity of demand is likely to face a higher price because, for that group, in the determination of **turnover** the lower level of sales will be more than offset by the higher unit price.

Two further points are worth noting. First, the mark-up price is not market-clearing, in the sense that there is no guarantee that it will ensure normal capacity utilization. Secondly, the mark-up price is neither purely a long- nor a short-term price but a hybrid. It is a short-term price in so far as it is drawn from the short-term supply function. However, it is long term in that the size of the mark-up reflects long-term views that excess profits may lead to civil riots or a threat to the catchment monopoly position from new competition. Moreover, this year's short-term function is derived from long-term decisions taken in the past to install new capacity. The present has to be seen as the end-point of a process of evolutionary development. This is the only basis on which to reconcile short- and long-term analysis.

4.4 A social service

The investigation of price determination here began by using what Max Weber called an ideal type – an idealized model of human interaction. The first ideal type considered here was a free market. The second type is a social service. In this case, the supply of fresh water is not regarded as a free market commodity, but as a social service in a country where central government is committed to state ownership of the principal utilities and where it is deemed a fundamental right of households to have access to water without paying a price for the resource. Each regional water organization is a statutory corporation, not a private company. Water for households is not metered. Farmers and industry are charged a price per m^3 equal to the short-term average prime cost at 80 per cent capacity. When consumption threatens to exceed supply, a supply fix is sought (Hirschleifer et al. 1960). Infrastructural investment is paid for by the water corporations borrowing from central government. The debt charge on these loans, as well as water's prime costs, are paid for from sales to agriculture and industry and from a fixed charge levied by the corporation on households. Typically, households receive a hidden cross-subsidy from the other two sectors.

In the real world, "actually existing" institutions can be found somewhere along the spectrum between free market and social service, and they take many, many forms. Some characteristics of the commodity approach are that water service companies are in the private sector; as private monopolies they are regulated by the state; there is no cross-subsidy between user groups; consumption is metered; and turnover is generated by **tariffs** derived from a mark-up on average prime costs. Characteristics of the service approach are that water service companies are government-owned; households' consumption is neither priced nor metered, but fixed charges are levied to defray the water service companies' operating losses; other users are required to pay average-cost-based tariffs on measured volumetric consumption; users with political clout are cross-subsidized; supply-fix philosophies predominate.

A hypothetical example of an intermediate form, with French practice in mind, is one where water is considered too essential to human existence and too evidently a natural monopoly to commit it into the hands of the free market. At the same time, there is no political conviction that supply must be handled through a state corporation (Kinnersley 1994: 5–6, 56, 170). In this case, political control is delegated to an elected regional government which, in turn, concedes a franchise to a regional private monopoly for water supply. The company collects water charges, and finances investment. But the real assets of the supply infrastructure are retained by the regional council. Customers' interests are represented by elected councillors, who grant and renew (or not) the franchise. They take an interest not merely in prices set, but the methods and frequency of billing, as well as payment arrangements for domestic consumers. The regional council recognizes the force of arguments put forward by environmentalists that unlimited demand for water imposes great environmental damage. It also understands how costly water is to produce in narrowly economic terms. For these reasons, all water consumers are metered and the private water supplier's cash outflow on prime and overhead costs, and its net profits, are met from its income from sales to households, farmers and industrialists.

With these two ideal types in place and an example given of an intermediate form, we can go on to consider the full range of issues that fall under the rubric of "demand management".

4.5 Approaches to demand management

It should be clear by now, that the concept of demand for water is a many-headed beast. Demand can be understood to refer to the needs of households, agriculture and industry; or to indicate the amount of water actually consumed; or, with the economists, to effective demand, that is, the price–quantity function in respect of the purchase of water. All three interpretations are perfectly valid, but, because they mean different things, it is important to be clear which one is employed; in this book it is the third sense above which is used, unless otherwise stated.

An important debate in the water industry in recent years has concerned consumption forecasting, usually referred to as "demand forecasts". The approach to forecasting in the 1960s and thereafter has been strongly criticized, as have supply-fix philosophies. The work of Rees with Williams (1993) is valuable here.

Objective forecasting techniques take two forms (McDonald & Kay 1988: 103–107). With *extrapolation* forecasts, past levels of aggregate consumption are recorded and their rate of growth is projected into the future. With *component* forecasts (also known as the causal or analytical technique), the different categories of water consumption are identified and their rates of growth into the future are estimated using demographic and economic projections relevant to each category. In this second case, there can be heavy data requirements. Perhaps for this reason, extrapolation is more commonly used.

The critique of forecasting method is wide-ranging, particularly in respect of extrapolation. The arguments are that it ignores the components of consumption, assumes linear growth into the future, often based on the steepest sections of the trend line, and adds overgenerous safety margins against unpredicted consumption surges. Component forecasts are innocent of the first of these charges, since this technique bases itself on population growth and predicted industrial and commercial development.

However, both techniques are guilty of two further methodological errors. One is the mis-identification of supply losses as a component of consumption, an issue already discussed in Chapter 2. The second error is to regard consumption as a simple technical issue, exogenous to other social changes. It is precisely these changes that are at the heart of what has become known as demand management. Demand management has at least five strands: internal and external re-use, consumption technology, land-use planning, educational initiatives and water pricing.

Internal and external re-use have already been defined in the text accompanying Figure 2.1. With internal re-use, a specific consumer first uses the water supplied to it from a water service company, then returns its wastewater internally for a second round of use, and then perhaps a third, and so on. For example, the installation of water storage during housing construction can make the one-off re-use of bath and shower water for flush-toilets perfectly feasible, reducing total household water use by perhaps 15 per cent.

With external re-use, a consumer, whether household, farmer or industrialist, uses its water supply, and then the wastewater is supplied as an input to another institution. With both internal and external re-use, the wastewater often requires treatment. Re-use can be seen as a supply-side innovation. At the same time, it brings about a lower aggregate consumption of water by the community as a whole than would happen in the absence of re-use, and, in that sense, re-use can also be regarded as a form of demand management.

Consumption technology, the second form of demand management, is similar to internal re-use, in that both deploy technical measures that reduce consumption of the external water supply. In this case, in respect of households, the redesign of showers, toilet cisterns, washing machines and dishwashers can cut water consumption significantly. Household access to the new technology may be through the purchase of new machines or in **retrofit** programmes of the type launched in some US municipalities. Household acceptance of such innovations can be underpinned by government through building regulations and water by-laws. In some countries, households are required by government to install rainwater catchment tanks, which is a supply-side intervention rather than demand management in the strict sense.

In the case of irrigation, McCann & Appleton (1993: 38) have pointed out that in France, for example, private abstraction is permitted by land ownership and past practice. It is a prime source of water wastage because of distribution losses, inefficient spraying devices and methods, and just straightforward over-spraying. The wasteful use of water in industry is also widespread in Europe, in the view of informed industrial opinion (Hills 1995).

Land-use planning is the third form of demand management. The argument concerns catchments where consumption is pushing at the very limits of supply capacity, or threatens soon to do so, and where increased abstraction would impose relatively high economic or environmental costs. In such cases, as Gordon (1993) and Merrett (1995) argue, land-use planning can restrain urban development – and the consumption that accompanies it – and divert its location to other regions not facing the same supply/consumption imbalances.

The fourth form of demand management is to use a variety of *educational measures* to persuade the population, as citizens, farmers and managers in industry, to use water wisely. Many US municipalities, for example, provide comprehensive consumption advice and information services, including residential water audits. "Trained personnel inspect water use fittings and appliances in individual homes, undertake leak detection and repair, evaluate lawn watering practices, advise on low-water garden design and recommend water-saving plants." (Rees with Williams 1993: 51)

US studies suggest that engineering audits, technical workshops, and best practice manuals can save 20–35 per cent of total annual water use with a short average payback period. Educational measures bring about a change in consumers' tastes and habits, which contribute to the conditions of effective demand already referred to in §4.2. This can be represented graphically as a leftward shift in the demand curve. Most orthodox economics takes preference functions as fixed (Hodgson 1994c).

A fifth form of demand management is *water pricing*, bringing us back to the effective demand for water. The metering of water use, and setting a price per unit quantity consumed, helps to underpin all the first four demand management forms. Water pricing should therefore be considered as complementary with these other measures, not as a substitute for them.

In many parts of the urban world, no price is paid for water. In the case of households, the reason can be found in traditions of the social service approach. In this case it is accepted that the supply of water should be regarded as a municipal service, partially financed by a local water tax. Wasteful use, stimulated by the zero price, is then built into consumption forecasts of the extrapolative and component type.

However, the growth in the economic costs of treating potable water, the environmental costs of abstraction, and the privatization of the water industry, have all brought metering and pricing onto the political agenda. In the UK in 1993, for example, 98 per cent of households lived in accommodation where their water consumption was unmetered and where the costs of water were met not by a price for its use but by a property-based fixed charge levied by the private water-service companies. The key difference in revenue raising between pricing and a fixed charge is that only the former is based on the quantity consumed by the household.

The other case where water consumption is unpriced is when farmers or industry have control over their own water supply and therefore provide it "free" to themselves. Of course, these users bear the economic costs of abstraction and treatment, but the zero price per unit is likely to lead to wasteful use. Moreover, farmers and firms do not directly bear the environmental costs of their abstraction.

Where prices are to be introduced for consumers, price-setting policy has to be considered. A theory of the ruling price level in free-market situations has been set out in Figure 4.4 and the accompanying text. In the case of institutions intermediate between the free market and a social service, water service companies' prices are also likely to be based on the short-term cost function, with an acceptable gross margin. The size of the margin is likely to be shaped by the investment plans of the catchment monopoly, for example, in raising water quality. In real life situations of regulated distribution, where infrastructures are owned by central or state governments, water for irrigation is often subsidized. Pakistan and California are examples.

Water price tariffs come in many varieties. There is likely to be a fixed standing charge for connection to the service, irrespective of total consumption. The unit price of water consumed may be on a flat rate basis, or increase (or fall) with each successive block of consumption. Higher prices at consumption peaks may be put in place. The peaking of water use has already been referred to in Chapter 2. It may be seasonal or in the so-called "needle peaks" at peak hours. User peaking, to be satisfied on the supply side, requires a greater system capacity than in the absence of peaks, and "average day in peak week" forecasts are conventionally employed in water investment planning. So, demand-side peak smoothing, in a long-term perspective, can offer considerable cost savings. For this reason, higher seasonal and needle tariffs are particularly attractive.

The discussion of tariffs brings us to important questions of the distributional effect of water pricing. Here, the concept of the **income elasticity of demand** is useful, a parallel idea to that of price elasticity. It is measured as follows: one records the total income and the total quantity purchased of a commodity for house-holds in two groups, A and B, where the income of group A is greater than the income of group B; the proportion by which A's purchases exceed those of B (the usual case) is calculated and divided by the proportion by which the income of A exceeds that of B; the ratio is the income elasticity of demand. Under stable con-ditions of demand, the concept can also apply to income *change* over time for a sin-gle household. Income elasticities are usually positive, unlike the price elasticity of demand.

It is widely assumed that the necessities of life have income elasticities less than one; that is, lower income groups spend proportionately more on these necessities than do higher income groups. Since water is the most basic of all needs, a move to water pricing may have a sharper effect on the poor. For those of us who are com-mitted to social justice, but also believe in the pricing of water, the tariff implica-tions are clear: there should be no standing charge for water use by households, tariffs should be of the increasing block variety, and the charge per unit within the first block – sufficient to provide for the basic needs of washing, cleaning, cooking and sanitation – should be modest.

The relationship between income and the consumption of water can also be seen at the macroeconomic level. For example, during the UK recession of 1989–93 there was a fall in water use in the industrial sectors, immediately giving the lie to extrapolative forecasting. A fall in the rate of housebuilding and in the purchase of water-based consumer durables may also have cut back the rate of growth in house-hold consumption.

4.6 Wastewater services

Thus far, in this chapter, the entire focus has been on the abstraction, treatment, dis-tribution and consumption of fresh water. But one should not neglect the collec-tion, treatment and disposal of wastewater. As McCann & Appleton (1993: 94) write:

The poor performance of all types of privately owned wastewater treatment plants has been a long-term problem for water quality regulators. Ongoing maintenance and operational supervision is often not given the attention it deserves, industrial managers being primarily concerned with the efficient operation of their main production processes.

Wastewater services are a **public good**, which economists define as any good where the benefit derived from it by one consumer does not diminish the benefit derived by consumers in general. The standard textbook example is a country's armed forces (Sandmo 1987: 1061–1067). Since no consumer privately appropriates a public good, it is impossible to price or to sell it.

Thus, the free market, untrammelled by government intervention, cannot establish wastewater service provision as a profitable activity. The *laissez-faire* economy is inherently incapable of dealing with pollution. The only exceptions to this rule in the world of water are the cases of internal and external re-use in Figure 2.1, where the benefits of wastewater treatment may, indeed, be captured privately. In principle, the social service institution would provide wastewater services for the common good, although the quality of service provision would be constrained by multiple claims on public finance from all the principal utilities.

In reality, as has been said, provision conforms to neither ideal type. Intermediate arrangements in the specific case where private or state-owned companies exist alongside powerful, democratic, central and regional government institutions, should have no difficulty in the provision of this public good. Several complementary measures are instituted.

First, in the domestic sector, collection, treatment and disposal is carried through by the water company. Since freshwater and wastewater functions are combined in a single catchment industry, and as fresh water is priced, the price to domestic consumers can be set to cover the costs both of freshwater supply and wastewater disposal.

Secondly, companies and corporations in the industrial sector are required to operate under a regulatory regime whereby they provide appropriate treatment for their own wastewater. The release of residual effluents into ground- and surface-water sources and the salt sea requires prior consent from government as well as discharge payments. Where industrial organizations generate effluents no different in general quality from households, as can be the case with offices and shops, they are permitted to discharge directly into the foul water system. In this case, water-charging principles are similar to those for households. However, where specific industrial pollutants are discharged to foul sewers, this would be agreed by licence with the water service company for both quantity and quality. A price formula is then used which is based on the additional costs imposed on the sewage treatment works to treat this extra pollution. Kinnersley writes (1994: 18):

> . . . one modern trend . . . is to apply charges for using the natural capacity of river basins, for discharging wastes as well as abstracting "raw" water. In

effect, this combination of legal and economic measures holds out possibilities of bringing access to river basin capacity into a more constructive relationship with a market economy. But the river basin itself cannot be a conventional market place, because transactions cannot be separated from each other, nor can interactions between each user be completely controlled.

Thirdly, agriculture is *sui generis*. Industrialized farming, such as the processing of calves or of turkeys, produces wastewater just as a manufacturing enterprise does. With arable farming, and livestock raised in the field, the water source may be precipitation or irrigation. In neither case is there a piped wastewater flow. Water consumption is either locked into the farmer's output, or seeps into the soil, or is lost in evapotranspiration.

But note that in agriculture and industry, discharge permits and charges do not deal adequately with the diffuse ambient pollution of water, as happens with acid rain or the nitrate contamination of groundwater from the use of fertilizers. For example, more cadmium enters the Rhine from road washings contaminated by car and lorry traffic than from point pollution.

For intermediate institutions, there are still two barriers to the satisfactory provision of wastewater services. First, it remains true that no-one privately appropriates this public good, and so companies will always be tempted to deliver as limited a wastewater service as they can get away with inside the regulatory framework. Limited provision brings lower treatment costs to the industry, and the burden is borne either by the public, which may have a weakly organized environmentalist movement, or by the silent world of mammals, birds, plants, fish and other creeping things, which neither pay taxes nor vote in the election of governments. Secondly, substantial improvement in wastewater services may require large capital investment, and regional government may be unwilling to accept the price increases for fresh- and wastewater services that new infrastructure would require.

For some years now, there has been a growing interest in incentive charging on wastewater effluents (RCEP 1992; Kinnersley 1994: 160–64). This has been stimulated by German, Dutch and French experience. All institutions discharging wastewater into rivers and lakes must do so within the constraints of the licensing system. A charge is levied which recovers the full cost of the discharge consent system. But the fees are set high enough to fulfil two other objectives. First, wastewater is still a polluting input into water courses, and such pollution imposes environmental and financial costs on other river and lake users, as well as additional treatment costs for downstream abstractors. These negative externalities can be reflected in the discharge fee, adopting the "polluter pays" principle, and can encourage dischargers to raise the quality of their wastewater, if the incremental costs of doing so fall short of the reduction in discharge fees as effluent quality rises. This is a clear case where marginal analysis is appropriate. Secondly, the surplus income from such fees can be used as a sourcing point for payments to users to introduce processes less polluting of water in the first place.

4.7 Case study: a Latvian water tariff

The Republic of Latvia is located in northern Europe, on the east Baltic coast (Fig. 4.5). Its geographical area is 63 700 km^2 and in 1995 its population was some 2.5 million, a ratio of 39 persons per km^2. From the time of Peter the Great at the beginning of the eighteenth century, until the Gorbachev era in the 1990s, it had been part of the Russian Empire, with the exception of two decades between the First and Second World Wars. Latvia achieved independence in August 1991. In April 1995, the rate of exchange of the dollar to the lats, the national currency, was 1:0.51. The lats is divided into 100 santimes.

During the first half of the 1990s, the macroeconomic situation was extremely unfavourable. In part this was attributable to the partial disintegration of Latvia's economic interrelationship with the Soviet economy, as the Soviet Empire collapsed, and in part it was attributable to government policies to maintain high interest rates and an overvalued currency. Reports from the Economist Intelligence Unit (EIU 1994, 1995a) showed that the growth rate of Latvian GDP had been negative in the five calendar years 1990–94. Throughout the same period, population declined. An estimated 80 per cent of the population lived below the poverty line.

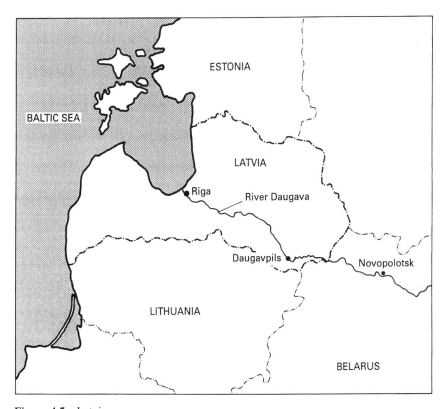

Figure 4.5 Latvia.

Real GDP per head fell in each of the three years 1991–3. Unemployment rose to the highest level of the three Baltic states (Estonia, Latvia and Lithuania) and many were on short-time working and on unpaid leave, as well as employees contracted to work part-time.

The city of Daugavpils is the second largest urban area in the country, lying some 220 km to the southeast of the capital, Riga. The city had a population in 1995 of about 120 000 people. It lies on the country's greatest river, the Daugava, which is 1030 km long and flows into Latvia from neighbouring Belarus.

In 1995, Daugavpils drew 70 per cent of its water supply from the river and a further 30 per cent from groundwater wells. The quality of the water from the river gave cause for concern. Novopolotsk, 100 km up stream in Belarus, was said to be a major source of pollution, particularly in respect of oil spills and discharge of heavy metals. In respect of wastewater, effluent from the wastewater treatment plant in Daugavpils did not meet the standards of the Helsinki Commission, a body set up to attempt to raise the water quality of the Baltic Sea. The Daugava leaves the city to flow to Riga, Latvia's capital city, where it joins the Baltic. Within the Baltic Sea environment programme, Daugavpils had been identified as one of the environmental hot spots.

A variety of investment options were under consideration for the city's water and wastewater utility, Udensvads, which was controlled by the municipal council. The economic and financial evaluation of these technical and managerial options all turned, at least in part, on the tariff to be set on Udensvads freshwater supplies and its sewerage services. These tariffs were central in determining the utility's annual turnover.

I was brought in as a consultant to prepare a tariff study, and argued that it was first necessary to agree on the objectives of any specific tariff regime. These were four: first, tariffs for fresh- and wastewater services must be affordable to the households purchasing them; secondly, tariffs should fully cover the prime and overhead costs, including debt charges, of service production, thereby eliminating the dependence of the water enterprise on local or national government subsidy; thirdly, tariffs should underpin the quest for sustainability by encouraging water conservation; fourthly, tariffs should encourage the protection of the environment from pollution.

At the time of the study, the level and structure of tariffs was set by the municipality, on the advice of the enterprise's management. In terms of number of customers, not only did individual households make up the great majority of individual connections, but total consumption itself was about 75 per cent from the domestic sector, with the rest deriving from local industrial and other customers, including those within the municipal sector, such as schools and hospitals.

The bulk of households were not metered. With metered customers one can confidently refer to the tariff as a price per unit of service. With those customers whose consumption is not known, the tariff in effect takes the form of a fixed charge based on some calculation of estimated consumption. It was necessary to consider whether or not major efforts should be made to meter those customers who were unmetered. With respect to households, these were overwhelmingly families living in flats in the tower and slab blocks to be found everywhere in the city.

In the household sector, it was not possible to recommend that a metering programme should be pursued in the short or medium term. The cost to a customer of meter installation had been calculated as equal to a month's average wage. At 1995 income levels, this was seen as too onerous a private burden. Of course, it was feasible in technical terms for Udensvads itself to fund meter installation out of its own budget, but the financial rate of return to such action would certainly have been negative, and for that reason could not be pursued immediately. However, metering was recommended for all industrial and institutional customers which had not yet been metered, as well as for domestic customers in all newly constructed housing.

An unmetered supply was in breach of objective three above, which encourages water conservation. However, metering of an entire block of flats, rather than its individual units, was perfectly feasible and it was argued by some that block meter readings, with charge per flat based on block consumption, could provide the foundation for the community in each block to commit itself to more frugal patterns of use.

The water tariff level which had applied locally from mid-1994 was 4 santimes per m^3 for domestic and municipal customers and 9 santimes per m^3 for industrial customers. In the case of wastewater services, the respective rates were 3 santimes per m^3 and 6 santimes per m^3. These were based on average prime costs at the time, equal to about 11 santimes per m^3 for fresh- and wastewater costs combined. In effect, the industrial sector was cross-subsidizing the domestic and municipal sectors. Moreover, there was no element of overhead cost in the cost calculation. This was because, under the arrangements developed during the era when Latvia was a constituent republic of the USSR, fresh- and wastewater utilities did not have to bear their own capital costs. This arrangement was still in force in 1995.

A recommendation was made that, from 1996, average total cost should be the basis of the tariff for both industrial and municipal customers, based on objective two cited above. If this regime were adopted, then as new investment programmes took place, with their associated debt charges, the price of water would rise.

With reference to objective four, it was recommended that a review be carried out of the quality of the effluent discharged into the sewerage system by Udensvads's non-domestic customers. Where this effluent was particularly harmful (e.g. to biological measures for sewage treatment, as with heavy metal discharges), a punitive wastewater rate should be levied. This would act as a powerful pricing signal for pre-treatment investment by industry, one of the most cost-effective of measures designed for environmental protection.

In respect of domestic customers, it was also recommended that the price of water be set at its average total cost. But in this case there was an important proviso drawn from objective one of the tariff regime: the price was not to be set at a level that would make customers' payments for the utility's services unaffordable. So, there was to be an affordability constraint imposed on the ATC rule.

The affordability criterion was derived from World Bank recommendations. The Bank has suggested that the combined tariff for fresh- and wastewater services should be set such that monthly payments by households should not exceed 3 per cent of average household income. Alternatively, the tariff should be set such that monthly payments do not exceed 5 per cent of the average income of the thirtieth **percentile**, counted up from the poorest individuals, of total income distribution. The second criterion has the strength of specifically recognizing that water and sanitation services should be affordable to those on relatively low incomes.

Unfortunately, no trustworthy data existed for Daugavpils, either on average household income or on the income of the thirtieth percentile. An alternative procedure was to calculate the average wage or salary of the city's working population and to multiply it by the average number of workers per household. Unfortunately, no data of this type existed either, although official statistics on average *full-time* salaries in the *public* sector were available.

The approach finally adopted was to create a synthetic household. The actual calculations are set out in Table 4.2. Census data was used to divide the population into three groups: those aged under 15, those aged between 15 and the official retirement age, and those above the retirement age. These three age groups were then divided into subcategories reflecting their socio-economic status, particularly with respect to the local labour market. The proportion within each category was then estimated on the basis of informed local opinion. The **modal income** of full-time workers and part-time workers, and the rate of pension, unemployment and child benefits, were estimated on the same basis. As Table 4.2 shows, this then produced an income flow of a synthetic individual representing the whole local population. That figure in its turn was multiplied by the average number of persons per household.

The strength of this approach was that it recognized the socio-economic heterogeneity of the local population, particularly at a time when macroeconomic conditions had created much underemployment, short-time working and unpaid leave. The method's fragility is that it relies heavily on local informed opinion for critical data inputs and is therefore as strong or as weak as the accuracy of such opinion. In retrospect, it would have been preferable to use a combination of data from informed local opinion, official statistics and, for example, the known wage and salary structure of Udensvads itself. This may have given a higher income estimate than the one chosen. What the true level of average household income in Daugavpils in April 1995 was, no-one will ever know.

The income calculated was used to specify the combined tariff for fresh- and wastewater services, using the following equation:

$$t = \frac{y \times w}{p \times c \times 30.42} \tag{4.4}$$

where t is the combined tariff per cubic metre of water supplied, y is the estimate of average household income per month, w is the World Bank's affordability criterion, p is the average number of persons per household among the utility's

Table 4.2 An estimate of average household income in Daugavpils.

Demographic category	% of population	Monthly earnings/benefits	Column 2 × column 3 (in lats per month)
Persons under 15	**20**	5	1
Persons aged 15–55/59, of which:	**60**		
In full-time education	7	5	0.21
Working full-time	50	45	13.5
Working part-time	11	28	1.85
Working short-time	6	28	1.01
On unpaid leave	6	0	0
Registered unemployed	8	20	0.96
Unregistered unemployed	2	0	0
Retired early	2	25	0.3
Incapacitated	3	20	0.36
Not seeking work	5	0	0
Persons above 55/59, of which:	**20**		
Fully-employed	3	45	0.27
Employed part-time	2	28	0.11
Not earning	95	25	4.75
Total	**100**		**24.32**

Average household with 2.4 persons at 24.32 lats per month gives 58 lats.

customers, c is the estimated water consumption in cubic metres per person per day, and 30.42 is the average number of days in a calendar month.

This gave the following result for 1995:

$$t = \frac{58 \times 0.03}{2.4 \times 0.25 \times 30.42} = 0.095 \text{ lats} = 9.5 \text{ santimes} \tag{4.5}$$

This, of course, was recommended as the upper limit to the combined tariff, set on the affordability grounds of objective one. Below that constraint, the average total cost rule of objective two could apply. However, the affordability limit was *already* exceeded by average operating costs, let alone the average total costs that might accompany an investment programme.

What becomes clear, when one examines the market of a city-based utility for fresh- and wastewater services, is that its customers all lie within a few kilometres of the plant itself. The effect is that the capacity of the market to meet tariffs for domestic, industrial and other institutional consumption is entirely determined *by the state of the local economy*. Daugavpils had been hit particularly hard by the economic recession that began in 1990, and the official local unemployment rate in 1995 was 10.6 per cent compared with the national average of 6.7 per cent. In these conditions, the ability of an investment programme to demonstrate financial viability would turn on the amount of capital spending in the investment programme that would be met from grant payments (e.g. from the richer Baltic countries) and the recovery of the local economy from the recession.

4.8 Case study: the Middle East and the virtual water thesis

Lundqvist et al. state (1985: 1):

> Land and water constitute two of the basic resources in the life support system on which man depends. With increasing standards of living, growing population and the geographical concentration of human activities, the pressure on these resources is being intensified at an alarming rate in large parts of the world. There is rapidly growing concern about the environmental hazards associated with this development if it is allowed to continue unchecked. The concern is related both to a depleting resource base as well as a deprivation of the long-term carrying capacity of our environment.

There is nowhere in the world to which these words apply more strikingly than the countries of the Middle East. In this case study, Jordan is the main point of reference in an account of the pressures for policy changes.[1]

Lundqvist's reference above to "a depleting resource base" is well illustrated in the case of Jordan. Schiffler et al. (1994: 2–4) write that it is among the countries with the scarcest water resources in the world. In flow terms, these amount to $170 \, m^3$ per year of renewable water per capita. This is far less, for example, than Syria, Egypt, the Lebanon and Turkey. Annual average precipitation is 8500 million m^3, of which 92 per cent is lost to evapotranspiration, 5 per cent goes to surface water and 3 per cent infiltrates the earth to recharge groundwater.

Fresh groundwater resources include **fossil aquifers** without any recharge. The largest of these is located in the Disi area, and one estimate suggests its exhaustion in about 50 years. With respect to renewable aquifers, the safe yield is the amount that can be abstracted each year without endangering groundwater supply in the future. Jordan's total annual safe yield from renewable aquifers is 275 million m^3.

In respect of surface water, the country's two main rivers are the Jordan and the Yarmouk, as well as **wadis** flowing to the Jordan, the Dead Sea and the Wadi Araba. Heavy Israeli pumping deprives the country of the use of the Jordan, and Syrian use of the Yarmouk limits Jordan's annual abstraction from the latter to 100 million m^3. The remaining surface-water resources add up to some 430 million m^3 per year. Schiffler et al. state (1994): "Almost all of this surface water will be used in the near future, when some ongoing dam projects will be completed".

---

1. This draws heavily on the work both of Professor Tony Allan of London's School of Oriental and African Studies, and of Manuel Schiffler and his colleagues at the German Development Institute (GDI) in Berlin (Schiffler et al. 1994, Allan 1995a, Allan & Karshenas 1995).

In 1991 (the most recent year for which full data are available) 833 million m^3 of fresh water were used: 74 per cent by agriculture, 21 per cent by municipalities (including households) and only 5 per cent by industry. Since the safe yield of renewable aquifers, plus the *total* surface-water flow (let alone an environmentally sustainable rate of surface water abstraction) were less than actual annual abstraction, Jordan clearly suffers from an ecological deficit in water. Note that the available data permit the calculation neither of self-sufficiency nor of catchment stress, as defined in §2.8.

Allan (1995a), in a paper on the political economy of water in the Middle East, has presented a scheme of the historical forms of water supply since 3000 BC. This is reproduced in Table 4.3. In his supply-side analysis with Massoud Karshenas (Allan & Karshenas 1995), they attack the "hydroparanoia" of academics, journalists and government officials of the region, who have mistakenly predicted conflict over water resources during the past two decades. Allan has also sought an expla-

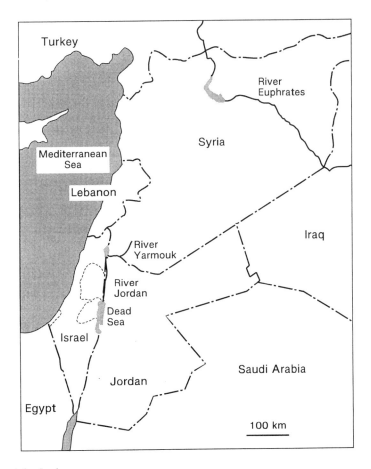

Figure 4.6 Jordan.

nation for the protraction of the transition in the Middle East from supply-fix policies to demand management. His response is the virtual water thesis, of which an exposition is presented below.

Table 4.3 Historical form of water supply in the Middle East.

Period	Surface water	Groundwater
3000 BC to 1970	Flood recession systems	Shallow wells, qanat/falaj* systems
1900 to present	Minor storage works – First Aswan Dam	
1930 to present		Deep wells to 50m
1960–70	Major storage works – Aswan Dam Major water carriers – Israel	
1970–2000	Major storage works – Turkey Major water carriers – Syria and Turkey	Very deep wells – to 500m Libya and Saudi Arabia

Source: Allan (1995a: table 1).
* **qanat**: The system used in Iran for moving water from an aquifer at the foot of a mountain to a point some considerable distance into the plains; the equivalent Omani term is **falaj**.

Since the second World War, there has been a strong long-term growth in international trade in staple grains, such as corn, rice and wheat. The global context of this development – the roots of which can be traced back to the nineteenth century – is complex. It includes the institutional pressure from farmers in North America and western Europe for the government subsidy of higher output volumes; technical change in grain-growing practices, such as the introduction of new crop species appropriate to large uptake of the macronutrients nitrogen, phosphorous and potassium; high population growth rates in the developing world unmatched by equivalent expansion in agricultural output; and the use of food aid as a tool of foreign policy. Specifically with reference to the Middle East, per capita cereal production fell in 1951–92, apart from the rise in the second half of the 1980s as a result of Saudi production, which has been significantly reduced since 1994 (Merrett 1971, Le Heron 1993, Dyson 1994; Allan 1995a: 41).

For any tonne of a grain species exported from the fields of a specific country, it is relatively easy to quantify in tonnes the amount of water used in agriculture to produce that tonne of grain, or of any other crop. Some of this water is trapped as H_2O molecules in the grain itself; some is locked in the plants from which the grain is removed; some infiltrates the soil to add to the groundwater resource; by far the greatest part is lost to the hydrosocial cycle through the processes of evapotranspiration. Allan has used the term "virtual water" to refer to the water used to grow exported crops (Allan 1995a, Allan & Karshenas 1995).

The water content of grain is, then, tiny in comparison with the total water necessary to grow it. This total, virtual water, has been conceptually invisible in the analysis of global trade in agriculture. However, in respect of the Middle East, understanding the import of virtual water is critical to understanding the region's ability to grow in terms of GDP and population, despite the aridity of its climate. It is the import of virtual water which has made wars over water unnecessary, which has permitted the long dominance of supply-fix measures, and which has constrained the space available for a demand management philosophy. The first shift away from the supply-fix position was in Israel in the late 1980s, consequent upon two successive droughts, as well as US pressure in the political wrestling over its continued subsidy of the Israeli economy.

Allan concludes (1995a: 42):

> The message . . . is that the region's governments have been able to take a less than urgent approach to managing their water according to sound economic and environmental principles because there has been a ready supply of extremely cheap water available in a very effective and operational system – world trade in food staples. Because of the measures being urged by the World Trade Organization and by influential elements of the food surplus producing industrialized economies, the circumstances of the past cannot be projected into the next decades. Also, there is a possibility that some pivotal food importers could exert unprecedented demands on the global food exporting system. As a result, the supply of cheap and secure water may not be so readily available in future and further there may be progressive price increases to add further discomfort.

In my view, Allan's work provides a powerful, creative metaphor which identifies the crux in the Middle East's use of water in recent decades. Below are set out some reflections on the virtual water thesis.

In the first place, it is essential to recognize that propositions concerning the *supply* of virtual water to the Middle East, or the *import* of virtual water, are metaphorical, not literal truths. It is not, of course, the case that Egypt, for example, both imports grain and imports the water consumed in growing that grain. Egypt imports the end product, not its inputs.

Secondly, expressions such as "the supply of virtual water" suggest we are dealing here with an invisible supply-side development. But Allan's metaphor can be set out in prosaic terms which suggest a different perspective. Here is the argument. The ability of Middle Eastern countries with arid climates to grow in terms of GDP, of population and of total food consumption, has been possible largely for two reasons: first, conventional supply-fix engineering, such as major storage works, major water carriers and very deep wells, have increased the supply of fresh water (see Table 4.3); secondly, the bulk import of staple grains has made it possible to consume food on a much expanded scale, yet the use of water by local agriculture has been far less than would have been the case had such levels of food consumption been supplied entirely from within the Middle Eastern agricultural sector.

In brief, there has been a sectoral adjustment of the Middle East economies, reducing the proportion of food consumption met from local food production, and this fall in dependence on domestic food has brought with it a *comparative* fall in local water use, although only in Israel has there been an *absolute* fall in consumption.

Here, the exposition stresses conventional supply-fix policy alongside sectoral adjustment in water use. In this perspective, the demand-side shift in recent Middle Eastern history begins, not with the Israeli policy reformulation of the 1980s but with the growth in Arab and Israeli bulk food imports.

Thirdly, and once again drawing inspiration from the virtual water thesis, human geographers may find it fruitful, in understanding the global space economy since 1945, to explore the global patterns of dependence on other nations' freshwater supplies which have paralleled global trade in food. If, metaphorically, virtual water is imported by country A from country B, then it follows that virtual water is exported from country B to country A. A country-based or region-based set of water accounts, drawn up as a matrix of virtual water export and import, would be the basic analytical tool here.

Let us again return to Jordan. An ecological deficit in water exists, as has already been argued. Supply-fix continues, with dam projects that will virtually complete the country's use of its entire surface water resource. Allan and Schiffler (ibid.) also report a growing interest in demand management, not least because of the high growth rate in Jordan's population.

Schiffler et al. clearly believe that, even in the mid-1990s, Jordanian GDP growth is constrained by current allocations between users of scarce water. Moreover, the conflict between economic development and water supplies appears set to become worse. New supply fixes are technically feasible but expensive. The re-use and recycling of water in the industrial sector are promising, but their impact will be small, given that industry's share of water consumption is so small. Household use is only some 85 litres pppd and a reduction here is indefensible. Agricultural use can be made more efficient through technological innovations and a shift in cropping patterns. But, in combination, these demand-side measures appear insufficient to the task.

The German Development Institute then takes the argument forward by introducing the concept of water productivity (Schiffler et al. 1994: 26–40). In principle, water productivity can be calculated for any sector or subsector of the economy and is equal to a sector's production of goods and services, measured in value-added terms per unit period of time, divided by water used measured in m³ per unit period of time. Value-added is a sector's contribution to GDP and can be calculated at factor cost or market prices. It is the sales revenue of a sector less purchased goods and services such as materials, intermediate products, spares, power and insurance. Thus:

$$M_i = \frac{N_i}{R_i} \tag{4.6}$$

where M_i is the water productivity of sector i, N_i is the sector's output in value-added terms, for example per month, and R_i is the sector's monthly consumption of water in m³.

Water productivity estimates are then produced for Jordan's economic sectors and subsectors (see Table 4.4). Schiffler et al. suggest (1994: 22):

Water productivity is an indicator for the economic value of water in a productive activity. It indicates as well the opportunity cost arising if water is not any more used in that activity. It can help, thus, to assess the impact of instruments for water demand management.

Table 4.4 shows water productivity in industry is 40 times higher than that of agriculture. On this basis, the GDI asserts "in the long term – assuming population growth and rising living standards – Jordan has no choice but to promote industry and services at the expense of irrigated agriculture" (ibid.: 28).

The heart of the matter, then, is that the supply of fresh water in the Jordanian hydrosocial cycle is said to constitute a physical constraint on the growth of the economy. This appears to be true, although the GDI never explores the technical feasibility and cost of leakage reduction as a means of expanding supply. So – the argument continues – the degree to which different sectors draw down this scarce resource, relative to value-added output, is a most useful indicator in policy formulation. This is absolutely correct, particularly with demand management policies.

However, the water productivity variable *as it is set out in Equation 4.6* is an inappropriate and misleading measure of a subsector's drawdown of the water resource. This is because it measures only the water flow *into* the subsector, for use in the production process, but neglects to take account the water flow *returned* to the hydrosocial cycle. This return flow is quantifiable as the sum of the external reuse and recycling loops of Figure 2.1.

The shortfall of the return flow with respect to the inflow is the water lost to the hydrosocial cycle. It is composed, first, of water molecules locked into the product, for example in soft drink manufacture; and secondly, of water losses through evapotranspiration, for example in agriculture.

So, it is worthwhile making a broad distinction – as do engineers – between consumptive use, where the return flow is small; and non-consumptive use, where the return flow is high. This point was already addressed in §2.8, and gives an alternative definition of water productivity:

$$M_i = \frac{N_i}{\alpha R_i} \tag{4.7}$$

where α is the water lost to the hydrosocial cycle (inflow less return flow) divided by the inflow. α lies between 1 and 0. At a value of 1 we have the case of Equation 4.6. At a value of 0, water productivity is infinitely high, for the production process does not draw down the scarce resource at all. At a value of 0.5, M_i in Equation 4.7 is twice the unadjusted value of Equation 4.6.

Using this correcting factor as in Equation 4.7, water productivity is a useful indicator for demand management. In the case of Jordan, it is likely to strengthen

Table 4.4 Water productivities in sectors and subsectors of the Jordanian economy in 1990.

Sector and subsector	M^*
Industry	
Paints	981.00
Pharmaceuticals	372.00–891.00
Household appliances	210.00–874.00
Diapers, toilet paper	681.00
Poultry slaughtering	181.00
Plastic foil	121.00
Brewing	85.00
Pharmaceutical capsules	82.00
Portland cement	74.00
Woven fabrics	67.00
Detergents	58.00
Petroleum refining	53.00
Soft drinks	48.00
Chocolate, biscuits, ice cream	48.00
Dyed fabrics	46.00
Thermal power plant	43.00
Steel bars	34.00
White cement	31.00
Tanning	30.00
Distilling	24.00
Steel pipes	10.00
Phosphate mining and purification	8.00
Potash	7.00
Noodles	6.00
Yeast cultivation	6.00
Phosphatic fertilizers	6.00
Tomato processing	2.00
Industry average	11.20
Agriculture	
Grapes	0.52
Potatoes	0.50
Beans	0.48
Citrus fruits	0.25
Bananas	0.18
Aubergines	0.18
Cabbage	0.07
Broad beans	0.04
Wheat**	0.01
Agriculture average	0.28

Source: Schiffler et al. (1994: tables 4–6).
*In Jordanian dinars per cubic metre. 1 JD = 1.5US$ in 1993.
**More precisely, in this case M equals 0.004!

the GDI's case for a planned sectoral adjustment from agriculture to industry, an adjustment that has long been under way through the import of virtual water.

However, note that both Equations 4.6 and 4.7 refer only to quantities of water used, not its quality. In particular, Equation 4.7 draws on the Chapter 2 concepts of re-use and recycling, which, again, are quantitative measures. In considering the appropriateness of re-use and recycling, it is essential to recognize that the quality of returned water is different from the water when it was originally abstracted. Schiffler (pers. comm.) notes that industrial wastewater in Jordan, because of inadequate treatment, is "often not a good for the environment or for agriculture . . . but a bad. Crop losses, and probably also health hazards, have occurred in Jordan as a result of the re-use of insufficiently treated wastewater in agriculture". The specific case of Jordan illustrates the general case that quantitative policy for the return of water to the hydrosocial cycle must be complemented by qualitative policies in respect of wastewater treatment.

4.9 Final remarks

This chapter began by distinguishing three concepts from each other: the varied needs that water meets, the quantity of water consumed in any time-period, and the effective demand for water. The basis of effective demand is the willingness and ability of households, farmers and industry (broadly defined) to purchase water in defined amounts. When the conditions of demand are stable, the price–quantity relationship can be represented as a mathematical function. From this, the price elasticity of demand can be derived; that is, the proportionate change in quantity purchased divided by the proportionate change in price. It was hypothesized that the effective demand equation is cubic.

Next, price determination is considered using two ideal types of social organization. In the free market, it is assumed that the price-setting procedure of management takes estimated average prime cost for the year ahead at normal capacity utilization and applies to that a percentage mark-up. Differential pricing, where it is feasible, sets a higher price for customer groups with a relatively low price elasticity of demand, or where supply costs are relatively high for that group.

In the case of supply by a company working on social service principles, it is assumed that there is a zero price and no metering for households. Industry and agriculture *is* metered and faces a price set equal to average prime cost at normal capacity utilization. The company's total supply costs are met by the sales revenue from industry and agriculture, plus a fixed charge levied on households, plus government subsidy. In practice, water service companies are to be found somewhere along the spectrum between free market and social service.

With the analysis of effective demand in place, a brief review is given of what is universally known as "demand management", although, in the language of this book, "use management" or "consumption management" would be more appro-

priate. Demand management is composed of at least five elements:
- internal and external re-use
- consumption technology
- land-use planning
- educational initiatives
- water pricing.

All are promising developments in contrast to the supply-fix dominance of the past, and all five are complementary to one another. The outline at the end of Chapter 2 of thirteen forms of supply option and, in this chapter, five use options (those of demand management), show how broad is the range of measures available for water resource planning.

Wastewater services, a public good, are considered next. In the real-life institutions that lie between the free market and social service ideal types, the supply of wastewater services to high standards, whether by private companies or publicly owned ones, requires a strong environmental regulator and substantial capital financing.

The Latvian case study, based on the city of Daugavpils, is drawn from a country in the difficult transition from that of a command economy integrated within the Soviet Empire to a market economy renewing its political and economic links with western Europe. In 1995, the city's water utility provided a clear example of a social service enterprise. The development of a new tariff for fresh- and wastewater services had simultaneously to address the costs of production, demand-management principles, environmental concerns over pollution, and the affordability of the tariff to local households. A consultant recommended a new tariff set at the average total cost of service provision, a cost that would rise, of course, with any new investments made to raise potable water standards and to increase the quality of treated wastewater returned to the local river. But it was also recommended that the costs borne by households through a fixed charge should conform to World Bank affordability guidelines. The estimation of local household income used the synthetic household approach, but, because of the paucity of good data on local incomes, it produced a highly speculative calculation. The result suggested that, in the absence of subsidy, it would not be possible to cover average total cost, in the eventuality of new investment, *and* conform to the affordability criterion. The poverty or prosperity of a local economy is shown to be critical to the effective demand for a water service company's outputs.

The Jordanian case study is of an arid country exhibiting strong quantitative consumption pressures on its limited water resources. The dominant user of water is agriculture. Allan's virtual water thesis is examined and is argued to be a powerful and creative metaphor that points to one of the salient features of modern Middle Eastern economic history. This is that a sectoral adjustment has taken place in the economy such that the proportion of food consumption met from local food production has fallen, bringing with it a fall in water use, in comparison with the situation were countries such as Jordan to be self-sufficient in food. The case study also examined Schiffler's use of the concept of water productivity and argued that it is a valuable analytical tool provided the measure is adjusted to recognize the return flow of water consumed to the hydrosocial cycle through external re-use and recycling.

One now has in place the analysis of both supply and effective demand, producer expenditures and user payments. So, the point has arrived when it is appropriate to consider the evaluation of projects, in terms of their costs and their benefits.

CHAPTER FIVE

Social cost–benefit analysis for water projects

5.1 Introduction

Political economy is a policy-orientated discipline in which the significance of economic inquiry is a function of its relevance for problem solving, including those dilemmas faced by governments. Virtually every central government has some overview on how it sees its distinct economic sectors, such as that of water, developing over time. Within sectors, programmes will be elaborated, and programmes require projects, the cutting edge of economic change.

The provision of government finance for a country's river-basin, water-supply and sanitation activities is always likely to face a budget constraint. It follows that, in choosing one set of initiatives demanding government resources, others inevitably have to be rejected or at least postponed. For example, Kinnersley (1994: xviii) has referred to the competition between irrigation projects and the provision of basic water supplies in some countries. The question then arises, from the point of view of the country as a whole, what is the optimum combination of initiatives to select in any given year? An entire branch of economics, known as social cost–benefit analysis (SCBA), can assist in answering this type of question. SCBA concerns itself with the evaluation of capital projects. SCBA and its application to water projects is the theme of this chapter. The political economy approach to SCBA is "the application of the theory of collective action to public finance". There is no assumption that the technique is a substitute for politics; rather, it interacts with politics (Schmid 1994: 105).

The chapter begins with a brief neutral exposition of the basic method. It then continues with issues specific to SCBA that the economist must confront: the choice of the discount rate, the use of shadow prices, the exclusion of transfer entries, and various estimation problems. Thereafter, the ground shifts to consider whether project analysis can have any relevance to demand management. Lastly, the chapter ends with two case studies and final remarks.

5.2 The project

The characteristics of a project are that it will usually have a well defined geographical location; it is an activity with a specific starting point at which time investment costs are incurred; and during the period of the project's life a series of outputs are produced which in most cases are capable of identification, measurement and valuation in money terms. The lining of a set of irrigation canals is an example. Gittinger notes (1982: 5): "Usually [the project] is a unique activity noticeably different from preceding, similar investments, and it is likely to be different from succeeding ones, not a routine segment of an ongoing program."

The project cycle can be classified in alternative ways. Here is an example that highlights the process as an economic activity:

- *Identification*   In this stage the project is conceived and its broad outlines agreed. This might come about as a result of a sector survey of broader scope.
- *Preparation and analysis*   A more detailed homing-in on project definition occurs with feasibility studies of alternative approaches, covering economic, financial, technical and organizational aspects.
- Ex ante *evaluation*   The project is evaluated in economic terms prior to its launch, on the basis of forecast expenditure and income on capital and current account. *Ex ante* social cost–benefit analysis is appropriate at this stage.
- *Implementation*   This breaks down into the gestation period, when the bulk of the investment costs are incurred, and subsequently the working period of the project until it comes to an end.
- Ex post *evaluation*   This takes place after the gestation period and can be prepared during the project's productive life or at its end. In the second case, all the data are retrospective, in contrast with the forecast and prospective intelligence of *ex ante* evaluation. The value of *ex post* appraisal is in developing subsequent strategies, programmes and projects on the basis of the wisdom of hindsight.

The general objectives of evaluation in economic terms are to test whether the returns on projects exceed their costs; to rank projects with respect to their economic efficiency; and to help policy-makers choose a set of projects, within the financial constraints facing them, which contribute most effectively to the country's economic development in a manner consistent with its social, political and environmental values. In the context of this book, SCBA is a technique primarily intended to raise the efficiency of government expenditure in support of water infrastructure programmes.

In its preparation and analysis phase, the project cycle should include an appraisal of the impact of an investment proposal on specific actors, such as the inhabitants of an area to be flooded for dam construction purposes, the civil engineering industry, regional government, farmers and manufacturing firms. This is vital to understand who is likely to benefit or to lose from the project and what their anticipated behavioural response may be. But the method for the evaluation of capital projects concerns itself primarily with the costs and benefits of the activity from the point of view of the nation as a whole. For this reason, project costs and returns are considered in terms of the real economy, that is, the flow of real resources used up in order to produce flows of real outputs or services. This holistic rather than partial approach is why the term "social cost–benefit analysis" is used.

5.3 The net present value approach

A starting point for understanding the technique of *ex ante* SCBA is to see it as a means to judge whether any single project will (or will not) make a positive contribution to a country's GDP. This requires that, over the full life of the project, that is, the gestation period plus the working period, one identifies the annual real flow of goods and services produced with the project in place, in comparison with the real flow of output without the project. This will be called the incremental real flow. At the same time we identify the incremental annual real flow of resources used in order to produce that output.

These identified physical quantities of outputs and inputs, in the form of goods or services, can then be given a common standard of measurement, in terms of their market prices. A spreadsheet is drawn up of the annual value of the incremental real output, in row 1, and the value of the incremental real costs in row 2. These two rows are hereafter referred to as the project's benefit and cost flows. The difference between benefit and cost in each year is the net benefit flow and is recorded in row 3. All this has been done in Table 5.1 for a water project deemed to have a full life of 25 years, starting at the beginning of year zero and terminating at the end of year 24.

Table 5.1 The net benefit spreadsheet.

	Year	0	1	2	. . .	22	23	24	Total
1	Value of incremental real output	b_0	b_1	b_2	$b(\ldots)$	b_{22}	b_{23}	b_{24}	B
2	Value of incremental real costs	c_0	c_1	c_2	$c(\ldots)$	c_{22}	c_{23}	c_{24}	C
3 (1–2)	Net benefit flow	n_0	n_1	n_2	$n(\ldots)$	n_{22}	n_{23}	n_{24}	N

Where the project has clear external effects in GDP terms beyond, so to speak, its own perimeter, and where these are quantifiable in price terms, such positive or negative externalities should also be included in the benefit, cost and net benefit flows. To be more precise, external GDP effects are those that arise for persons and institutions other than those engaged in the project's gestation, its production processes and the use of its outputs.

Kinnersley has pointed to the frequency of negative externalities from water projects (1994: 17):

In economic terms, the most common hazard is the way in which rivers spread damage caused at a particular point. Activities that pollute the river or divert part of its flow usually affect water quality or flow for some distance down stream. The water may thus be made unsuitable for other uses down stream, but often without would-be users down stream being able to trace who or what caused the pollution and thereby claim compensation. Thus, the polluting upstream activity, assuming it is a business, may be said to be escaping costs of waste disposal and preventive measures against pollution which it should normally have to provide for and reflect in its selling prices.

In the early period of a water infrastructure project, the net benefit flow is likely to be negative, when capital account outlays fail to be matched by the stream of benefits. Thereafter, net benefits may remain positive through to the end of the project's life, although major rehabilitation outlays in the mid-years could modify this general rule. In all cases we would expect the identification and valuation of benefits to be even more difficult than that of costs; this is the only general law of SCBA.

An important issue arises immediately. Should the market prices we are using for the valuation of the incremental real flows be constant or **out-turn prices**? **Constant prices** are those ruling in any specified year and are applied to the out-comes of both that year and all other years considered in the analysis of the project. Out-turn prices are the ruling level of prices in each separate year and, of course, vary over time. The standard practice in the field is to use constant (usually base-year) prices. The argument is that a general price inflation does not, in comparison with zero inflation, indicate any difference in the real value to society of outputs or inputs. This rule notwithstanding, the planning of a project's financing should consider the likely level of general price inflation, so that an adequate budget is obtained to cover the costs of the gestation period.

However, if there is a strong expectation of a *relative* shift in the prices of some outputs or inputs, quite aside from any general movement in the price index, these would be reflected in the calculation of the net benefit stream. Here the assumption is that relative change implies a shift in the value that society, through the market, places on the commodities produced, or consumed in order to produce.

A second issue that deserves to be addressed at an early stage concerns the mode of project finance. Here the question is: Should SCBA as a general practice embody in its calculations the sources of the money that will fund the project during its gestation period? Gittinger argues persuasively that this should not be the case (1982: 46). He suggests that one assumes all financing for a project comes from domestic sources and that all its returns go to domestic residents. This is consistent with the general objective of SCBA defined in terms of projects' contributions to GDP. On this basis alone, the decision is made as to whether or not a particular scheme should be approved. This convention, splitting the evaluation of individual projects from their specific financing sources, is almost universally accepted by project analysts. It is perfectly consistent with identification of possible financing sources in the preparation and analysis phase of the project cycle that precedes or runs alongside *ex ante* appraisal.

By this stage, we should have completed a spreadsheet such as Table 5.1. Let us suppose that N is positive, that is to say, the total value of the stream of incremental real output (B) exceeds the total value of the flow of incremental real costs (C). In such a case, is it correct to assume the project should be approved? Can SCBA really be that easy?

Unfortunately not, and this is for two distinct reasons. First, such a procedure would ignore the timing of the benefit and cost flows. Secondly, such a procedure would provide no basis for choosing between projects, either where two (or more) are mutually exclusive for technical reasons; or where a budget constraint imposes a limit on the total number of projects that can be financed – the starting point of this chapter. The term "project interdependence" will be used to refer to either of these two situations.

With respect to timing, in project evaluation it is universally accepted that each annual benefit and each annual cost entry should have its value reduced, that is, it should be discounted, the later it occurs. More formally, the proportionate reduction in the value of an entry between any two years should be the same. The reason why this view is held is discussed in §5.4. Here one deals simply with the arithmetic.

The mathematics of discounting are set out below. The accounting conventions are that the total project period is divided into years, that rates of growth or contraction are expressed in annual compound rates, and that all receipts and payments in each year occur at the beginning of that year.

Suppose one dollar is received at the beginning of year zero and that in each successive year the sum received increases at a constant compound rate of 5 per cent. Then, after five years this series of payments in dollars and cents is set out in Table 5.2.

Table 5.2 A compound growth series.

Year	0	1	2	3	4
1 Receipt ($)	1.00	$1.00\,(1.05)^1$	$1.00\,(1.05)^2$	$1.00\,(1.05)^3$	$1.00\,(1.05)^4$
2 Which equals ($)	1.00	1.05	1.1	1.16	1.22

Thus, the entry for each receipt in any given year t is equal to:

$$1.00\,(1.05)^t \tag{5.1}$$

More generally, for any given proportionate rate of increase i, expressed as a decimal, we can write the receipt as:

$$1.00\,(1+i)^t \tag{5.2}$$

This is a compound growth series. But when we discount the future, we are implicitly suggesting not growth into the future but contraction back from the future. We argue that an entry t years after the present (the start of year zero) needs to be reduced proportionately each year to calculate its present value.

So, analogous to the general expression for a growth series, in the reverse case of discounting, a receipt of one dollar in year t at a discount rate equal to d, has a present value which equals:

$$\frac{1.00}{(1+d)^t} \tag{5.3}$$

One divides, instead of multiplying, by the exponential term. As a double check, note that in the first row of Table 5.2, year 4's receipt when divided by $(1.05)^4$ gives a present value of 1.00.

Returning to Table 5.1, the decision to discount each and every entry after year zero should be carried out, as Equation 5.3 above indicates, through dividing it by $(1+d)^t$ where, again, d is the selected discount rate (such as 5 per cent) and t is the year in which the entry appears.

These individual discounted values of the benefit, cost and net benefit streams are added up and recorded. They will be written here as B^*, C^* and N^*. N^* is known as the **net present value** (NPV) of the project. Where the NPV is positive, the project can be approved on economic grounds, provided that it can be considered discretely, that is, in the absence of project interdependence.

The apparently laborious task of dividing dozens of cell entries by $(1+d)^t$ can be simplified with a calculator, or a published discounting table, or with an NPV function available through standard spreadsheet computer software. In the approach above, all costs and benefits in the first year are undiscounted, since they are assumed to occur at the beginning of year zero. (Note that $(1+d)^0 = 1$.) Alternatively (and this is World Bank practice) the first year of the project is counted as year 1 (not year 0), all entries are deemed incurred at the end of each year and, once again, we have the convenient expression $(1+d)^t$ to use to discount, where t is the project year in which the entry falls. Here, first-year receipts and outlays *are* discounted.

Next it is necessary to consider the difficulty that, in common experience, not all project proposals are appropriately considered in their own right, as independent of each other. In the first place, on technical grounds two (or more) proposals may be mutually exclusive, most obviously when each is an alternative means by which a water infrastructure investment can secure a given policy objective. In the case of mutually exclusive projects, the solution is simple: calculate the NPV of each option and choose the proposal with the highest NPV.

In the second place, projects may not be appropriately considered as independent because a single programme budget is available for their financing, and the budget is insufficient to fund them all. The question is then posed: On economic grounds what is the optimum subset of proposals to which finance should be provided? In this case, for every proposal (including all mutually exclusive projects) we calculate the **net benefit–investment ratio**. This is simply done by dividing the full project life into two: the gestation period, when the net benefit flow is consistently negative, and the working period when the net benefit is, in most years, positive. The NPV for each period is separately calculated and that of the working period divided by that of the gestation period. All projects are then ranked from top to bottom, using the net benefit investment ratio; lower-ranking mutually exclusive projects are eliminated; and then all the remaining projects are approved off the top until the budget constraint blocks any further approvals.

What we have done in the second case, that of financially interdependent choices, is to consider the key constraint as the budget funds available during the gestation period and then to rank all projects in terms of the productivity (in NPV terms) of these funds. This maximizes the total NPV generated by the use of the funds available. The net benefit investment ratio can also be considered as a means not of excluding projects but of ranking them in terms of their starting date. One hastens to add that such a ranking and selection procedure is a recommendation for the decision-making process, not a description of what always takes place, as one shall see.

5.4 Discounting the future

Up to this point, the net present value approach to the evaluation of capital projects has been presented in terms of its most basic elements. But if SCBA in this form is to be an acceptable technique for project decision-making, some extremely important areas of debate must be introduced. These are classified below under four headings: discounting, **shadow prices**, transfer entries, and estimation errors. Environmental costs and benefits are discussed in Chapter 8.

In §5.3 it was argued that the timing of the benefits and costs of a project should be handled in the evaluation process by reducing these values in proportion to the time delay before the benefit or the cost is registered. The method for doing this was described as the division of each value by $(1+d)^t$, where d is the selected discount rate and t is the year in which the entry appears. But no reason was given as to why the timing of benefits and costs is significant, nor why they should be handled by means of a discounting process.

The justification for this approach falls strictly not within the field of economics but of psychology. There seems to be an almost universal human propensity to regard a real benefit offered now as superior to the same real benefit offered later in time. Human life is beset with risks of injury, loss and death. That has always been the human condition and is doubtless the origin of a rather rational desire to receive now, when the probability of receipt is greatest, rather than later when such probability is perceived to be less. This mind-set is an example not of human frailty but of sweet wisdom distilled from the bitter lessons of experience.

Project evaluation, rightly in my view, recognizes this psychological propensity by the use of the discounting technique in respect of project outputs. This still leaves us with the difficult question: At what level should the time rate of discount be set? Since the evaluation of capital projects is here placed within a national framework, setting out from the concept of the **gross domestic product**, one is clearly seeking a value of *d* that is appropriate to society as a whole, and so one refers to this rate as the **social time rate of discount** (STRD).

To grasp the reasoning here, it may be of some help to draw a parallel with the physicist's concept of the half-life of radioactive material. In applying the arithmetic of discounting to handle the psychology of risk avoidance, we are suggesting that a real output (or cost) available now with a value of $1.00 has proportionately lower value today if available only in a year's time, and a still lower proportionate value in the following year, and so on and so forth. So, today's value of that specific real output contracts or decays at a constant rate, the later the time at which the benefit is registered, rather like the radioactive strength of some elements decays at a constant proportionate rate over time. For any given value of a real output (or cost) at the beginning of year zero, and for any specific measure of the social time rate of discount, it is easy to say how many years will elapse before that original value is deemed to have decayed by one-half. This number of years I shall call the half-life of the value of incremental real output implicit in any measure of the STRD.

In Figure 5.1 the half-life of the value of output, cost or net benefit is given for every integer value of the STRD between 1 per cent and 15 per cent. The results are dramatic. For a value of 1 per cent, the half-life is as long as 70 years; for 15 per cent, the half-life is only five years! The absolute value of the half-life falls substantially up to an STRD of about 5 or 6 per cent, where it averages close to 13 years.

This figure may help decision-makers set the appropriate rate for project evaluation in their own countries. In my view the whole thrust of discussions on sustainability in recent years suggests that high values of the STRD are inappropriate, as they produce short half-lives – they markedly devalue the medium- and long-term effects of project investment decisions. This point of view has been strongly urged by Lipton (1992) in the context of the high real rates of interest that were current in the 1980s and 1990s. A benchmark STRD of 3 per cent per year may be widely appropriate, which would give a half-life for net benefit entries of about 23 years. This half-life approach, set within an historical time perspective, is an antidote to the counter-intuitive "back-from-the-future" character of discounting (Bausor 1994: 327; Merrett 1989).

An alternative to the NPV approach is used by organizations such as the World Bank and most other international financing agencies, as Gittinger reports (1982: 331). This calculates the **internal rate of return** (IRR) of a project, that is, the discount rate at which the discounted value of incremental real output precisely equals the discounted value of incremental real costs. A cut-off rate is then set, and all independent projects above this value are deemed worthy of approval.

There are problems with this approach. First, the cut-off rate is said to represent the **opportunity cost of capital** in a country, but in fact no-one knows what this is. Moreover, the concept of the opportunity cost of capital rests on the assumption that its use would lead to the approval of all possible investments scoring above the cut-off rate. Yet such an outcome is entirely inappropriate, since choice on that basis ignores the huge importance of environmental and distributional effects in correct decision-making (see Ch. 8). Secondly, the IRR cut-off rate is used as a crude ranker of projects, but it is recognized that the IRR should not be used to rank projects at all (Gittinger 1982: 332). Thirdly, when constant prices are being used, chosen cut-off rates are typically in the range 8–15 per cent. If the appropriate STRD is much lower, as I have suggested, then the use of such a high cut-off rate for the IRR introduces a bias against sustainability, favouring projects with lower (but earlier) undiscounted returns.

The advantages of using a net present value approach based on the social time rate of discount are that it forces decision-makers to make choices about the appropriate social time discount rate in their society; that once this is set, no unjustifiable bias in favour of the short term takes place; and that it gives a practical ranking for interdependent projects, using the net benefit investment ratio, which recognizes in a pragmatic way the opportunity cost of capital funds.

Given that the IRR is used for project ranking, yet also discriminates in favour of the short run, its use by the World Bank (where the cut-off is almost invariably 10 per cent) and in seminal texts such as that of Gittinger (1982) is a major error in current SCBA practice and theory.

Before moving on, a few brief words are in order on SCBA's poor yet hard-worked cousin, social cost-effectiveness analysis (SCEA). Broadly speaking, whenever two or more projects achieve the same outcomes but in different ways, they are mutually

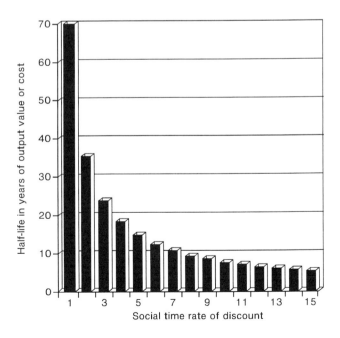

Figure 5.1 Project value half-lives and the STRD.

exclusive on technical grounds, as we have seen. Project analysis limited to such a package of alternatives can be carried through merely by calculating which option has the lowest discounted social costs, that is, which is the most cost-effective from the point of view of society as a whole. Winpenny (1994: 72), for example, cites the case of irrigation canal lining in southern California in the use of SCEA, and this technique will appear again in both of this chapter's case studies.

5.5 Shadow prices and transfer entries

In §5.3, market prices are used to convert incremental real outputs and real costs into the value terms of Table 5.1. Project evaluation should always do this for its first estimate of the NPV. However, there may be cases where the market price of an input or output is not deemed to be the best measure available to assess the project's contribution to gross domestic product. This is usually where the market price of an input does not measure its opportunity cost adequately. The opportunity cost of a real input to the production of goods or services is the output value forgone elsewhere in the national economy as a result of that input's use for the project under appraisal.

Shadow prices
Where a unit price other than the market price is used, so that the values of incremental outputs and costs are recalculated, we speak of shadow prices. In the text below are listed the principal situations where shadow prices are used. The topic is dealt with fully by Gittinger (1982: ch. 7).
  • In projects that use unskilled labour in a local labour market area where unemployment or underemployment is high, the shadow price for the wages of these workers should be set equal to zero. The United Nations calculates that as many as 30 per cent of the world's 2400 million workers are not "productively employed". In the case of interdependent projects, the shadow price procedure will strengthen the ranking position of initiatives which are intensive in their use of unskilled labour. The same argument applies to semi-skilled (or even skilled) labour, provided un- or underemployment is high in their case also. In general, a zero shadow-wage for specific groups of workers will disfavour water infrastructure projects, if they are capital intensive, in comparison with labour-intensive non-water projects. Of course, newly employed workers will receive a real market wage, not a zero shadow wage. As a result, their real income will rise, redistributing the annual flow of consumption goods and services. Some authors therefore argue that the shadow wage rate of the unemployed should be little different from their market wage, if workers have the temerity to spend their wages (Hughes 1991: 238–9). In fact, the zero shadow wage is justified on GDP opportunity-cost grounds. One can regard the distribution outcome as a bonus.

- The pricing of land will usually be at its market price, or an estimate of this based on its periodic rent. Where this price is believed to be distorted upwards as a result of local monopoly power, a lower price is appropriate based on land values derived from active rental markets. The use of a shadow price in this case tends to favour water utilities and river basin activities, since they are land-intensive. This is strikingly clear in the case of fresh- and wastewater treatment works with their filtration beds. Lamella technology, using parallel plates in water treatment, makes for more spatially compact plants and is now often used where small footprint infrastructures are required, such as in towns or scenic areas.
- Where the project's raw materials and manufactured inputs are purchased from industries operating at excess capacity, they should be valued only at their marginal cost, not their market price.
- Projects producing tradable goods (i.e. outputs that are actual or potential exports) must now be considered, as well as schemes using imported inputs. In many developing countries, the domestic currency is overvalued at its official rate. For example, the official rate of the dollar to the peso might be 1:1 rather than the market rate of 1:2. Whether the SCBA is carried out in dollars or pesos, overvaluation (which is often accompanied by tariffs on imports and subsidies to exports) tends to favour projects that are relatively intensive input importers and to disfavour those that are intensive output exporters. In such cases, SCBA would use a shadow rate of foreign exchange closer to the unofficial market rate. If a developing country or transitional country's currency is overvalued, projects that stimulate water-based tourism or which reduce any water imports would be strengthened by the use of a shadow price. However, water infrastructure projects that are dependent on imports of construction machinery, steel, cement, manufactured plant for treatment purposes, and on overseas consultants, will find their net present value cut sharply through the shadow pricing of foreign exchange. By way of extension, shadow pricing also would be used, to favourable effect, for project outputs that are import substitutes and, unfavourably, for project inputs that are diverted exports.
- Where the outputs of a scheme merely replace a river basin or utility service which is already provided, the value of the incremental real output should be shadow-priced at the marginal cost saving from no longer producing the good or service in its pre-project mode.
- Where a project's output is large relative to the existing volume of sales of that product or service, the market price is likely to fall. Conventional practice is to shadow price output half way between the market price with and the market price without the project.
- Where SCBA is applied to a project that brings an increase in the supply of water, clearly we need a water price in order to calculate the net benefit flow. Where no such price exists, for example because the payment for water is by a fixed charge, we face a real difficulty. One might enter a shadow price based on an average total cost calculation, but that would be unwise since it would give us

a cost–cost, not a cost–benefit, relationship. An acceptable alternative is to use a price taken from another catchment or country where the conditions of demand are broadly similar. But this is easier said than done. If that approach is impossible, it may be useful to set up several plausible shadow prices, and then carry out "what if?" calculations of the net present value. Finally, in desperation, one can convert the approach from SCBA to SCEA and estimate total discounted costs per cubic metre of water supplied over the project's total life. I used such an efficiency indicator to good effect in Armenia, in the autumn of 1996, to compare gravity-fed supply augmentation with the technical alternative of pumped supplies.

It would be absurd to expect the project analyst of any single investment to prepare, from a blank sheet, a set of appropriate shadow prices. These should be national parameters, and may be available from the Ministry of Finance.

Transfer entries
In the calculation of the NPV, the objective is to identify real outputs and real costs and then to value these at market (or at shadow) prices. In the financial accounts of a project, where project cashflows are the focus, certain transfer items usually appear. These include the capital sum of a loan, repayment of loan principal, payment of loan interest, receipt of government subsidy and payment of government tax. They represent only transfers of money between one actor and another, not the values of real outputs and inputs from the perspective of the national economy's GDP. So, in project evaluation, in contrast to cashflow accounting, these items would not be included in the net benefit stream of Table 5.1. Note that insurance payments are usually taken to represent a real cost, not a transfer payment.

In respect of these matters, Schmid (1994: 106), in his overview of cost–benefit analysis writes: "Without political interpretation it is impossible to tell if a tax, tariff or exchange control is a distorting mistake, to be corrected by a shadow price, or if it is implementing intended policy". There are two errors in this view. First, taxes and tariffs are transfer entries and are not liable to correction through shadow pricing. Secondly, local currency overvaluation through exchange controls is properly corrected through a shadow price, because this recognizes the scarcity of export earnings *vis-à-vis* the scale of imports. Exchange controls have a different goal – to strengthen the country's terms of trade – and shadow pricing does not undermine that goal.

5.6 Estimation errors

Ex ante evaluation makes an estimate of the project's NPV through forecasting future events: the length of the full project life, the real costs of producing real outputs, the future market prices of inputs and outputs, and the future value of shadow prices. Inevitably, *ex ante* analysis makes errors of estimation. When the project is

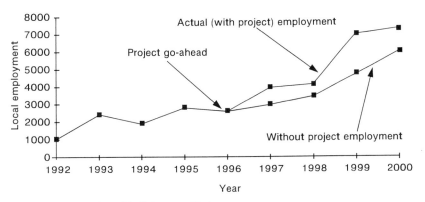

Figure 5.2   The unbearable lightness of being.

given its go-ahead, or is turned down, its NPV is a hazardous projection, not a fact.

Yet, *ex post* evaluation also faces substantial estimation problems, captured by Kundera in the wonderful phrase "the unbearable lightness of being" (Kundera 1984; see also Bausor 1994). This is illustrated in Figure 5.2. Let us suppose a major water infrastructure project is launched in 1996 in an urban area where inadequate sanitation exists alongside extensive unemployment. The justification for project approval includes positive reference to the expansion of employment in the local labour market area through construction work and the production of building materials. In the year 2000 an *ex post* evaluation is carried out. Actual employment in that year is compared, favourably, with actual employment in the year 1996 and the argument is made that the original project approval was, on employment grounds at least, fully justified.

This commits the most elementary error of project analysis. For the incremental approach adopted in §5.3 refers not to a change over time but to the difference in outcomes with and without the scheme. If, as Figure 5.2 suggests, the employment difference between situations with and without the project is much smaller than the actual changes in employment, because in any case employment was on an upward trend, the project's employment effects are much weaker than the 1996 to 2000 comparison indicates.

The unbearable lightness of being is that we can never know, in *ex post* evaluation, what would have occurred without the project in place, because that did not happen. We cannot rewind history to discover what would have happened if the project had been refused funding.

So, there is a curious symmetry between *ex ante* evaluation, which forecasts one unknowable future, and *ex post* evaluation which reflects on one unknowable past.

However, these evident difficulties should not be allowed to induce the paralysis of reason by the abandonment of analysis. Little & Mirrlees (1991: 356, 373) argue strongly that the net advantages of SCBA are considerable. Evaluation can be thought of as the partial elimination of error implicit in decisions by hunch or on grounds of apparent financial profitability. This suggests that NPV estimation procedures should focus on elements that seem likely to make a substantial difference to the outcome, rather than seeking greater accuracy in respect of trivia. Where powerful determinants of project success with a range of possible values are identified, sensitivity analysis can be used to test NPV response to different, plausible assumptions (Gittinger 1982: 363–71).

Just as important as the reduction of estimation error is the avoidance of estimation bias. In the assessment of project outcomes, specific water actors – in the private or the public sphere – will always have a material interest in specific projects receiving approval and others being rejected. The analyst must resist pressure to massage the figures.

A quite specific form of estimation bias, known as the McNamara effect, has unfortunately marked World Bank project appraisal for many years. When the supply of funds for projects is in full flood, pressures of an informal kind are placed on appraisers to inflate their calculation of the internal rate of return, so that more projects qualify for approval. Such manipulation can prove important for the internal promotion of the project analyst. Moreover, the analyst will have moved up, up and away should the project crash. In 1995, the *Financial Times* reported a growing number of problem projects because of this weakness in institutional culture (Graham 1995). Little & Mirrlees (1991) suggest that the McNamara effect is found not only in the World Bank but in much commercial bank lending to the Third World.

The quite extraordinary feature of World Bank project funding is that, although the IRR is always calculated prior to project approval and then re-estimated at the end of the gestation period using the project completion report, no further *ex post* evaluation is carried out. As a result, no substantial analysis by the World Bank exists on the success or failure in SCBA terms of the Bank's own projects over the years of their working lives (Little & Mirrlees 1991: 364–71).

A distinct case of SCBA misuse, rather than error of estimation, is where a specific water actor calls for SCBA to be carried out, knowing the time, cost and ambiguities within even the best such analyses, essentially in order to delay or block investment for some extraneous reason. In the UK, Ofwat has been charged precisely with this tactic, in order to hold back water-quality improvements (Kinnersley 1994: 175).

5.7 SCBA and the management of demand

Until this point in the chapter, the orientation has been to the supply-side of hydro-economics. This made sense because the chapter's focus is project analysis, and new abstraction investments, reservoirs, dams, fresh- and wastewater treatment plants and irrigation infrastructures are all meat and drink to SCBA. Yet, in the housing and residential infrastructure field, Malpezzi (1990) has demonstrated that SCBA can also be applied to initiatives that are not necessarily projects in the terms set out by Gittinger in §5.2.

This raises the possibility that the technique could also be deployed in appraising demand management activities. Chapter 4 distinguished at least five strands to such action: internal and external re-use, consumption technology, land-use planning, educational initiatives and water pricing. All of these measures, even if they cannot all be regarded as projects, certainly incur social (and private) costs: re-use consists of small-scale projects, requiring engineering work in agriculture and industry or building work in the household sector; consumption technology necessitates equipment purchases; land-use planning, at the very least, requires the labour and associated overheads of the planning profession; educational initiatives similarly demand important current expenditures, such as developing teaching materials and conducting awareness-raising campaigns; and water pricing requires metering as well as the staff to calculate and secure payment for individualized bills.

So, we can get a handle on the costs of demand management. It will be important to remember that, in their calculation, one should subtract any existing outgoings which are no longer necessary. For example, billing for metered water will eliminate the need to collect the fixed charge, so the net cost of the initiative will be lower than its gross cost.

But what about *benefits*? In the case where demand management is associated with an absolute fall in recorded water services supplied, gross domestic product will correspondingly decline. However, demand management in these cases will usually be directed to some environmental goal. So, the analysis will take the form of assessing demand management costs, including GDP losses, against environmental benefits.

However, in most cases it is likely that demand management leads not to a fall in water use but a slower rate of growth than would have otherwise occurred. In this case, we have two broad sets of options:

• no demand management, full supply growth, full environmental impact, full growth in water consumption
• demand management in place, limited supply growth, limited environmental impact, limited growth in water consumption

Each of these interdependent options could be separately evaluated using the procedures already laid out in this chapter. A simpler, but perfectly valid decision procedure, where either one route or the other is certain to be taken, would be to assess the situation by conducting a single SCBA evaluation – of the demand management option – by comparing:

• costs: demand management expenditures plus reduced water consumption
• benefits: reduced supply costs plus reduced environmental impact.

If the NPV is positive, demand management is to be preferred; if negative, the full supply option should be given the green light. What is certain is that the most intellectually challenging part of such a procedure would be the evaluation of "reduced water consumption" and "reduced environmental impact". In the latter case, we face the difficulties to be considered in Chapter 8. In the former case, we need to determine whether the services provided by water decline as a result of the shift to a lower quantity consumed. Often, in the case of technology change for example, no such fall in service provision occurs, and so "reduced water consumption" does not, in fact, constitute a cost.

SCBA/SCEA studies of demand management have appeared in recent years. For example, Hufschmidt et al. (1987) conducted a cost-effectiveness study of demand management versus full supply augmentation in the domestic and industrial sectors of Beijing, and identified the former as clearly more attractive. In contrast, Munasinghe (1990) investigated demand management practices in reducing calls upon a Manila aquifer to below its recharge rate, thereby saving on the long-term costs of securing new supplies. The two alternatives differed little in their evaluated outcomes.

5.8 Case study: a Peruvian simulation model

Peru is located on the western side of South America and has borders with Ecuador, Colombia, Brazil, Bolivia and Chile (Fig. 5.3). Its land area is $1\,285\,000\,km^2$ and in 1995 its population was 23.5 million, giving a density of 18 persons per km^2. The principal cities are Lima–Callao, Arequipa, Trujillo and Chiclayo. Seventy per cent of households live in urban areas. In general terms, the climate is temperate on the coast, tropical in the jungles and cool in the Andes mountain range. To the west of the highlands the climate is dry, but on the northern and eastern sides there is heavy rainfall between October and April. Of particular interest here is the 3000 km western seaboard, a narrow strip of desert, where the only fertile areas are some 50 valleys created by rivers flowing down from the Andes to the Pacific, where cotton, rice, sugar and many kinds of fruit are produced, and where there have been major investments in irrigation.

The government of President Alberto Fujimori, which came to power in 1990, is politically autocratic and economically neoliberal. From its inception, a strong commitment was made to the deregulation of the economy and the privatization of the 230 enterprises owned by the Peruvian state at the end of the 1980s. As the Economist Intelligence Unit reports: "(Fujimori) re-established relations with the international financial community and threw open the long-isolated economy. He slashed state spending, raised tax revenue, cut tariff barriers and passed new laws to encourage foreign investment" (EIU 1995b: 5). The 1993 Constitution enshrines the principle of the free market economy and reduces the role of the state.

At the time of the case study preparation in July 1995, the rate of exchange of the dollar to the Peruvian nuevo sol was 1:2.24. Bank and street rates were almost identical. The country's principal export is coca leaf and its derivative, cocaine paste. By mid-1995, privatization of state assets had been extensively implemented in the mining industry, hydrocarbons, electricity generation and distribution, manufacturing industry, hotels, banking, railways, ports, air transport and telecommunications.

In 1994 the Ministry of the Presidency of the Peruvian government sought the

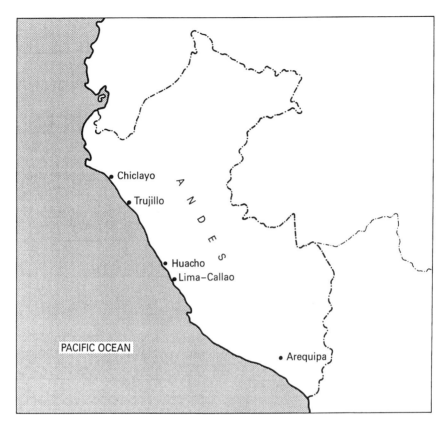

Figure 5.3 Peru.

assistance of the Inter-American Development Bank (IADB) in strengthening 10–20 utilities supplying fresh- and wastewater services in medium-size towns. Typically these were nominally private companies (Sociedades Anónimas), operating only at local district level and, in practice, controlled by the local municipality. Within the Ministry, the execution of this new policy initiative was assigned to a group entitled the Programa Nacional de Agua Potable y Alcantarillado (PRONAP).

PRONAP established a Subprogramme B principally aimed at identifying short-term projects totalling just a few million soles for each enterprise, which could rectify their base-period weaknesses of an institutional, commercial and operational character. The identification and evaluation of these projects was carried out in 1995 by private sector teams composed of at least one foreign and one domestic engineering consultancy. As the terms of reference for consultants state: "The sub-programme's basic objective is to assist the water utilities to transform themselves into enterprises which are well structured, autonomous and financially viable" (PRONAP 1994: 1).

The intention was that the Ministry of the Presidency, through PRONAP, would provide each enterprise with a grant to cover subprojects judged successfully to address base-period weaknesses. The sourcing of these ministerial grants was an IADB loan to the Peruvian government. Thereafter loan finance would be sought for medium- and long-term projects. Such loans would be made once again by the Inter-American Development Bank, but now directly to the reconstituted firms themselves.

Lima-based engineers suggested that this approach reflected a sea-change in IADB perspectives over the past ten years. In the past the Bank would be willing to provide loans for projects essentially defined in terms of their total costs and their physical character, knowing that borrowing by municipal and state bodies constituted sovereign debt guaranteed in the last resort by national government. However, the foreign debt crises of the 1980s and 1990s now produces a much keener interest in a utility's ability to repay its debts and a much stronger interest in the managerial, commercial and financial strengths of the enterprise. The global shift to privatization underpins this new perspective of the Bank, since these restructuring loans, as one might call them, can be the prelude to the sale of the municipal utility to the private sector. Here there is a marked similarity with the Latvian case study of Chapter 4.

The terms of reference supplied by PRONAP to the consulting firms bidding for Subprogramme B anticipated the use both of social cost–benefit analysis and social cost-effectiveness analysis in identifying the optimal set of restructuring subprojects. SCBA was to use the NPV approach with appropriate shadow prices supplied by the Ministry, and PRONAP specifically requested the use of the IADB's SIMOP model (Powers & Valencia 1978).

SIMOP (modelo de simulación de obras públicas) is a simulation model for the estimation of the costs and benefits of an expansion in the supply of drinking water to an urban population. It was written in Washington in 1978 by Terry Powers and Carlos Valencia of the IADB, and it consists of the software to run the simulations of production, distribution and consumption, as well as a user's manual. SIMOP is a paradigmatic case of the institutional production of what Hodgson calls a "congealed habit" (1994d: 304).

The manual begins with households' effective demand function for fresh water set within the neoclassical paradigm of economics, using the concept of water's "use value" in an inconsistent way from the very first page. The representation of the demand function is linear, in contrast with the cubic function proposed in Chapter 4. The supply function is represented as of zero elasticity in contrast to the quadratic function proposed in Chapter 3. The additional benefit to users of increasing the supply of water from one time-period to the next is then calculated, making an inadmissible use of the concept of consumers' surplus. The flaw here is the assumption that household expenditure on water would be an inconsequential proportion of household income. This assumption is a necessary condition of consumer surplus theorizing. But it does not hold true for water since, at high prices, families are forced to devote a considerable proportion of their income to water purchases because of the vital need for a minimum consumption. So, SIMOP's demand and supply analysis is fundamentally flawed from the outset.

Next, the important case is considered of a supply increase in one area for an in-migration of consumers leading to the release of resources in water supply in the area from which they migrated. Consideration is given to situations both where total consumption of the in-migrants falls and where it rises. By now it is clear that fixed costs are being excluded from the price analysis and later it becomes apparent that these will be covered by a fixed charge. The distributional implications of such an arrangement are not discussed. What is not made clear is that if price reflects only incremental prime costs and all other costs are borne by the standing charge, the very great bulk of turnover is likely to derive from the fixed charge, not the marginal price. This would make the IADB approach practically redundant (Powers & Valencia 1978: 2).

After these theoretical preliminaries, the remainder of the IADB text is concerned with the correct use of the model. We are informed that the analyst must choose between two types of demand curve: one linear, and the other curvilinear with constant price elasticity. It is demonstrated that, having made such a choice, then with the input of the existing price–quantity relationship and an estimation of price elasticity at the existing price, the entire effective demand function can be specified. The same section discusses the displacement of the demand function over time.

It now becomes clear that the representation of demand in the theoretical text as linear is not because that is considered to be its actual character, but to make it possible mathematically and computationally to construct an entire demand curve on the basis of the known current sales and current price plus the elasticity estimate. So, real economic relationships are falsely represented to create a working simulation model, rather than the model being designed to represent reality.

Table 1 of the manual provides an arithmetic example of the argument, for two hypothetical consumer groups receiving an increment in the supply of water from the expansion of a single urban network. Here two vital difficulties are encountered. The first is that with group one there is a fixed charge and a two-section, falling block tariff (see p. 66). Actual consumption is to be found in the second section. The question arises: What price should be used for the base-period price–quantity observation required in the model? In fact, the marginal price at the quantity actually consumed is used. This does not seem unreasonable.

The *second* difficulty is that group two faces a fixed charge and a rising block tariff of two sections, where actual consumption is to be found in the first section, for which the marginal price is zero and consequentially the price elasticity is not mathematically defined, because it requires division by zero. In fact, a zero-price situation is completely outside the scope of the simulation software and the theorizing that precedes the model-building. Pragmatically, the demand function is specified for this group by assuming it has the same slope as that for group one. Once again a fiction, with no socio-economic foundation, is created to make it possible to run the model.

The remainder of the manual includes discussion of allocation rules where future water shortages exist, types of incremental cost, shadow prices, investment decision criteria, optimal start dates and sensitivity analysis. An appendix provides guidance on the use of the software, written in the 1978 language of punch cards. A second appendix is a case study of the application of the model to the city of Medellín in Colombia in 1974.

In the 1995 case study, the first application of SIMOP was to Huacho, a town of about 125 000 persons on the western seaboard, 130 km north of Lima, Peru's capital.

The short-term investment programme proposed by the consultants for the water utility was composed of three segments: institutional, commercial and operational. There were three institutional subprojects. The first was intended to implant an enterprise management with a strong quality orientation, changing from departmental perspectives to a client/contractor outlook. The second sought to strengthen the capacity of management to think innovatively, to clarify the results it sought to achieve and to unify the motivation of the board of directors, senior management and the workforce. The third institutional subproject aimed at setting up new information systems which were integrated throughout the company.

Within the commercial segment, there was only one subproject, the various components of which aimed to achieve integration in commercial management. It included the setting up of consumption metering for industrial, commercial, cultural and domestic users, with a 50 per cent coverage target.

The operational segment contained three subprojects. The first was intended to rehabilitate the tube wells, guarantee chlorination of the freshwater supply, replace 12 per cent of the freshwater distribution network and rehabilitate the facilities for its control. The second sought to improve the sewerage system and raise the firm's capacity to maintain it by acquiring maintenance equipment, replacing 19 per cent of the system and cleaning another 21 per cent. The third subproject aimed at strengthening the operational functioning of the firm while promoting its technical cooperation with two other water utilities in the area to raise the quality control of the water supply and operational control of the system's networks.

A social cost-effectiveness analysis for these seven subprojects was carried out by calculating the costs of each in terms of discounted shadow prices, assigning a priority point score to each within the range of 1–20, and then calculating the present cost per point scored. By far the best performer on this measure was the second institutional subproject, with costs of only US$6000 per priority point. The poorest performer was the second operational subproject, with costs of US$84 000 per point. However, full confidence in the SCEA was limited by the fact that three different persons assigned the scores, one scorer for each segment, and no criteria were established for the scoring process. A useful account of scoring and weighting appraisal is provided in the UK Department of the Environment's 1991 publication *Policy appraisal and the environment*.

The consultant's economist (myself) next turned to social cost–benefit analysis. Of the seven subprojects, it was considered feasible to estimate the economic benefits of only two, the commercial initiative and operational subproject one. In the former case, metering was anticipated to reduce consumption by current and future users. In the latter case, system improvement seemed likely to offer significant leakage reduction, bringing a 4 per cent potential rise in fresh water distributed to customers in 1997, falling to 2.4 per cent in the year 2015.

PRONAP wished to see SIMOP employed to give a net present value calculation of the estimated costs of all seven subprojects against the benefits of the two where benefit estimation seemed plausible.

The fundamental problem facing the work lay on the effective demand side of the model. For any single class of users, its demand curve needed to be estimated for each of the 21 years from 1995–2015. To make this possible, SIMOP required three types of information: first, the quantity consumed and the marginal price paid in 1995, the year preceding the project launch; secondly, the elasticity of demand at that point on the 1995 demand function; thirdly, the annual rate at which the demand function would be displaced to the right after 1995.

However, in Huacho a survey had established that 97 per cent of domestic consumers either had no meter or had a meter that was faulty and could not be used for invoicing. Engineers were willing to make an estimate of the quantity of fresh water consumed by domestic consumers, although the likelihood of error was considerable here because the scale of unaccounted-for-water was unknown. However, since the costs of water supplied were met entirely by the fixed charge, no marginal price existed. Thus, the fundamental price–quantity relation for 1995 – on which the whole SIMOP exercise depended – did not exist. In fact, the absence of pricing for water was predominantly the case for domestic consumption throughout Peru in 1995.

Obstinately, the consulting economist made the heroic assumption that the appropriate marginal price in Huacho was equal to the ruling market price in the city of Chiclayo in 1990 – the only other Peruvian water price available. Thereafter, the price elasticity for demand was plucked from a range provided by PRONAP and the function's displacement was assumed equal to the forecast rate of growth of Huacho's population.

Now the work took on an Alice-in-Wonderland quality, for the assumed market price of fresh water of 0.2 soles per m^3 could not possibly be consistent with the estimated real level of water used at a time when water had no price. So, the real quantity consumed put into the database was reduced by a third to make allowance for this.

The simulation model was run and gave an NPV of –7.8 million soles. The reasons for this were clear. First, the costs of all the subprojects were included but only the estimated benefits from two of them. Secondly, the reduction in stated water consumption, done to be consistent with the fictional market price, meant that the Huacho plant apparently was producing at only 48 per cent of its capacity. So, there were virtually no gains from domestic metering or leakage reduction since, apparently, twice as much water was available as was consumed – this at a time when in Huacho, but not in Wonderland, large numbers of consumers suffered from imposed rationing. Note too that, on the demand side, functions for a 21-year period were estimated without a single observation of a price of water in Huacho or a quantity consumed.

5.9 Case study: the Pergau hydroelectric power project in Malaysia

This case study addresses the *ex ante* economic appraisal of the Pergau hydroelectric power plant, located on the Pergau river in the State of Kelantan in Peninsular Malaysia, close to the border with Thailand. The project's purpose is to contribute to Malaysia's electric power generation, specifically for peak load requirements of some four hours per day. The plant's capacity is 600 MW. Construction began in 1991 and was scheduled for completion in 1996. The documentary sources for the case study are: a report by the UK National Audit Office (NAO 1993); a report by the House of Commons Committee of Public Accounts (HoCCPA 1994); and a two-volume report of the House of Commons Foreign Affairs Committee (HoCFAC 1994: vol. II). Along with the Victoria dam in Sri Lanka, Pergau is the most extensively documented hydroelectric power project in the history of the British aid programme, for reasons that will become evident below.

A brief project history is in order. The site was first identified in the 1960s by the Malaysian authorities. In 1987, a World Bank power sector report on Malaysia noted the river as a potential site for a 211 MW station, providing base-load power, but argued that the country should concentrate entirely on gas-fired electricity generation until the year 2000. Then, in 1988 the Snowy Mountains Engineering Corporation (SMEC) of Australia, working on a higher projected rate of growth in power demand than the World Bank had, estimated the project cost for a 600 MW plant, providing peak-load requirements. At 1989 prices, this cost was £208 million.

In the autumn of 1988, the Malaysian electricity authority (Lembaga Letrik Negara) identified the Pergau site as a priority in planning for its future power requirements, and a British engineering company, Balfour Beatty, was encouraged to take an interest. The company formed a consortium with a second British company, Cementation International, jointly to pursue the matter. At the end of the same year, the consortium submitted a formal application for British aid support under the Aid and Trade Provision. ATP was first established in 1977 as a separate part of the aid programme, its purpose being to support sound developmental projects of particular commercial and industrial importance to the UK. The consortium's application was based on the SMEC feasibility study and it gave an indicative cost which, after rounding up, was £316 million, 52 per cent higher than the original SMEC estimate. The Malaysian government was keen to proceed, but this was dependent upon the availability of ATP support, which was first formally offered in April 1989.

In engineering terms, the scheme consists of an underground power station drawing water from the Kuala Yong dam located on the Pergau river. Flow capture is enhanced by the pumped transfer of water from adjoining southern catchments through a 24 km aqueduct rock tunnel. A second, re-regulating, dam controls the outflow from the power station (HOCFAC 1994: vol. II, 214). The scheme's salient features are: the Kuala Yong dam, 75 m in height with an active storage volume of 54 million m^3; the power station with four 150 MW vertical turbine generators in the main cavern and four 3-phase transformers in a separate hall; the reregulating dam, which prevents flood surge down stream; the aqueduct, its pump station and associated weirs and drop shafts; the power transmission and switchgear infrastructure, linking the project to the 275 kV national grid with additional 33 kV/11 kV local distribution lines; the power tunnels with three main sections: a 1 km low-pressure tunnel, a 470 m deep high-pressure shaft and a 3 km tailrace tunnel; and 68 km of access roads for construction and maintenance, through mountainous terrain.

Now the environmental issues are considered, which have been reviewed by Friends of the Earth in a lengthy memorandum (HOCFAC 1994: vol. II, 132–63). The Pergau river winds through heavily forested mountains and is shallow, narrow, heavily silted and boulder-strewn. There are no major industrial or population centres in the project area. At least 37 species of mammals, 140 species of birds, and innumerable insect species, as well as amphibians and reptiles are present. The mammals include the Sumatran rhinoceros, elephant, tapir, tiger, seladang, squirrel, macaque, bamboo and hollow-faced bat, civet and otter. The catchment area is mostly covered with evergreen tropical rainforest. A field survey identified 289 species of flowering plant, 98 species of fern and 51 species of moss. Environmental Resources Management, a British consultancy, suggested that biodiversity is above average with respect to the remaining Malay Peninsula rainforest. The area is used by the Orang Asli, who are aboriginal hunter–gatherers.

The project's impact would derive from the clearance of the site, the opening up of the new road network, the intrusion of site works, and the creation of a new settlement for 2000 construction workers. There might be negative impacts for the paddies tilled by farmers down stream, and the reservoir might create a new habitat for vectors and intermediate hosts of tropical diseases. The new road infrastructure might also be the basis for encroachment by new local settlements as well as increased logging. Already in 1991, the Kelantan State government was planning to develop a 400 hectare vegetable dye plantation in the reservoir catchment area, with Japanese aid assistance.

Environmental impact studies were made in 1988–91, of considerable variation in quality. This book is not the place to review them. On their evidence, the UK's Overseas Development Administration (ODA) came to the view that the project was acceptable on environmental criteria and posed no resettlement problems. In contrast, Friends of the Earth are strongly critical of the environmental analysis in the key 1989 document and suggest that, in any case, environmental evaluation played no part in the go-ahead decision for the project (HOCCPA 1994: 2).

The economic appraisal must now be considered. Until recently, infrastructure projects in water, transport and power have been dominated by a "predict and provide" philosophy. Demand forecasts of a methodologically naïve type are made and then a supply strategy is prepared to deliver those needs. This has been an engineer-driven approach. After all, the physical capacity to produce and distribute infrastructural services of various kinds is their business. The world we live in would be unimaginably different in their absence. In respect of water and transport, the predict and provide mind-set is now fast changing, as we saw in Chapter 4. In Europe, the phrase "predict and prevent" has even been coined. But in the late 1980s, supply-fix was still dominant in power generation. So, in Malaysia as elsewhere, the strategy was to prepare consumption estimates for a long-term planning period and then decide on the sequencing over time of appropriate projects. Economic appraisal consisted not of SCBA but of SCEA, where each means of electricity generation is compared in discounted cost terms with its alternatives, and an optimum sequence is derived. This was the case with the Pergau hydroelectric power project.

The appraisal of the project can be usefully classified into three time-periods: first, up to mid-March 1989; secondly, from the end of March through to April 1989; and thirdly, after April 1989 and through to July 1991.

Up to mid-March 1989

Aid and Trade Provision proposals at the time were usually identified by British firms and submitted to the UK's Department of Trade and Industry (DTI). If the Department was satisfied by the likely industrial and commercial benefits to the UK, the proposal was recommended to the ODA. For its approval, the proposal should have to satisfy the normal aid criteria, including economic considerations. The DTI told the ODA in October 1988 that Pergau was a likely candidate for ATP support, the consortium provisionally estimating project capital costs at £200–300 million (NAO 1993: 3–4).

Following an initial desk-study in October–November, ODA decided on a cautious approach because the project was large and only marginally viable. In January 1989, the consortium revised the capital cost estimate to £316 million. On this basis, ODA considered that the project's justification presented major problems in terms of size, price and timing. At the end of February, a minute from the ODA economist summarized advice from the World Bank to the effect that definitive economic appraisal required the ODA and Lembaga Letrik Negara to run a full system cost analysis of Malaysia's power sector with and without Pergau on the basis of revised demand forecasts and the project's capital costs.

A more substantial appraisal was carried out over a seven-day period in mid-March, including a two-day mission to Malaysia from 12 to 14 March by an ODA economist and a DTI officer, who worked with Lembaga Letrik Negara power engineers. An advance oral report to the British High Commission and a telephone call from the Commission to the ODA in London on 14 March resulted in Prime Minister Thatcher making an offer of aid of £68 million for Pergau to Prime Minister Mahathir on 15 March, conditional on a full economic appraisal.

The National Audit Office records (NAO 1993: 4):

The appraisal report was produced on 20 March. It concluded that there was a reasonable developmental case on the basis of the Snowy Mountains estimate of [£208 millions at 1989 prices]. Although recognizing that Balfour Beatty/Cementation International's costs were not yet firm, the appraisal noted that, at £316 million, they were more than 50 per cent higher on a direct comparison of equivalent constant prices. Taking account of partial offsetting savings from lower transmission costs, the appraisal concluded that the economic viability of the project at this price was marginal.

The ODA had wished to make the appraisal mission considerably earlier than mid-March, but met resistance from the consortium, the DTI, and the British High Commission in Kuala Lumpur, which argued that it would destabilize the commercial negotiations under way. Sir Tim Lankester, Permanent Secretary at the ODA, later stated that "commercial pressures at that time were very great indeed". The DTI's view was that it was highly desirable to give opportunities for the British power generation equipment industry in Malaysia. The industry was short of work at that time (HoCCPA 1994: viii, 6).

There were pressures too coming through the Prime Minister's Office as a result of industrial representations. Mr John Lippitt of GEC – the main subcontractor to the Balfour Beatty/Cementation joint venture – wrote to the PM's Private Secretary on 13 March:

> The difficulty in Mrs Thatcher giving a firm assurance that ATP support will be available to the Malaysian Prime Minister when they meet tomorrow arises from the fact that the ODA have suddenly decided they require more information about the project before they can come to a firm decision; this despite the fact that they had a full feasibility study (the SMEC study) well before Christmas and could have made the necessary enquiries well before now. In fact as a result of promptings from Balfour Beatty and ourselves, they have sent a team to Kuala Lumpur this weekend to do the work.

An enclosed note confirmed the total value of the project at approximately £300 million (HoCFAC 1994: vol. I, xxix, lxii).

End-March to April 1989
On 31 March, at a meeting with ODA officials, the consortium submitted a revised contract proposal of £397 million, taking account of adverse geological factors not recognized in earlier assessments, the additional cost of ancillary equipment and of all work to final commissioning. That is to say, within two weeks of Mrs Thatcher's first offer to Prime Minister Mahathir, Balfour Beatty and Cementation International had increased their capital cost estimate by 26 per cent (NAO 1993: 4). The ODA was "totally astonished" by this. Its distress was particularly acute because the January estimate had been presented as a maximum figure, as Balfour Beatty acknowledged at the 31 March meeting (HoCFAC 1994: vol. I, xxxi). The

Committee of Public Accounts stated: "We put it to the Administration that this increase in the price immediately after the agreement with the Prime Minister was a matter for serious concern and indeed suspicious" (HoCCPA 1994: viii). In all events, the ODA informed the consortium that the economic viability of the project had been just acceptable at £316 million, but was no longer so at £397 million.

In spite of this, on 17 April the UK government made a formal aid grant offer to Malaysia, open for six months, of £68 million as part of a mixed credit to finance the UK content of the earlier estimated contract price of £316 million. The government also indicated a willingness to consider an increased level of support, in view of the 31 March price rise.

May 1989 to July 1991
Over the next two years, the aid offer lapsed twice and was twice renewed. In January 1991, an agreed contract price was set at £417 million, the increase mainly reflecting the inclusion of Export Credits Guarantee Department's (ECGD) insurance premium and the taxation of expatriate staff. Economic appraisal work continued, by the ODA and Lembaga Letrik Negara, and by July 1991 the economic position can be summarized as follows below (NAO 1993: 5–7, HoCCPA 1994: x, xii, 5–6, HoCFAC 1994: vol. I, xxii):

- From an economic point of view, Pergau was unequivocally unsound.
- This choice of a hydroelectric power project would, in discounted terms, impose an additional cost burden on Malaysian consumers of £100 million over a 35-year period.
- The project should be delayed for many years or deferred indefinitely.
- In response to the view that hydroelectric schemes will last for a hundred years, the view was taken that this would not happen without further investment and that critical decisions on whether to extend the life of the scheme or close it down were likely to be necessary within 35 years at a maximum.
- The cost-effective alternative to Pergau for power generation was gas turbines. (This was primarily because Malaysia has massive reserves of natural gas with little alternative use.)
- The project had the potential to crowd out other projects from ATP support.
- Proceeding with the project would reduce the ECGD's ability to offer insurance cover in Malaysia as the review limit for exposure in the country would be exceeded by almost half.
- Far from aid contributing to the development of Malaysia, it would at best be offsetting the extra cost of choosing Pergau.
- There were no serious countervailing arguments against rejection. Moreover, there was local opposition to the project.

In February 1991, Sir Tim Lankester emphasized to Ministers that his responsibility was to ensure that aid funds were administered in a prudent and economical manner and he believed that providing funds for the Pergau project would be inconsistent with this, concluding that the project should not be supported. The decision, in fact, to go ahead with the project was taken by the Foreign Secretary,

Douglas Hurd, in the same month, in consultation with the then Prime Minister, John Major. On 4 July 1991, Hurd gave instructions to incur expenditure of up to £234 million against the aid programme over a period of 14 years. Lankester later called the project ". . . an abuse of the aid programme . . .". Made in public, these are strong words for a British civil servant (HoCCPA 1994: 16).

The increase in the estimated cash cost to the aid programme of 244 per cent between April 1989 and July 1991, from £68 million to £234 million (see above), was due in part to the escalation in project costs and in part to a decision to meet the aid commitment by **backloading** the cashflow in using a soft loan package rather than mixed credit funding, which "would have blown the ATP budget". But the back-loaded soft loan was more expensive in cash terms. As a result, Pergau involved the largest cash sum ever provided for a single scheme under the Aid and Trade Provision.

The controversy over the approval and financing of Pergau did not end with the reports of the National Audit Office and the House of Commons. In May 1994, the World Development Movement took legal action on the matter and in November of that year the High Court ruled the Foreign Secretary had acted unlawfully in earmarking aid funds. The Court indicated that subsidy for the project was fatally flawed because it was economically unsound and did not promote the development of a country's economy as required by the 1980 Overseas Development and Cooperation Act. What becomes evident from press reporting, as well as the Foreign Affairs Committee Report, is that the UK government in 1989 was using the aid programme as a bargaining counter in order to secure a £1300 million arms deal with Malaysia (*Financial Times* 1994: 11 November, 12 and 14 December; *The Observer* 1994: 6 February; HoCFAC 1994: vol. I, ix, x, xxxviii).

What do we learn from this hydroelectric power project with respect to economic appraisal? First, that when a "predict and provide" philosophy rules in power-generation planning, the appropriate technique is to select and implement projects in a sequence that matches electricity supplies to demand forecasts at the lowest total discounted cost. Secondly, that where commercial interests are strong and where a project is not subject to the normal tendering procedures, but arises from bilateral negotiations between client and contractor, intense pressure will be imposed on the evaluators to arrive quickly at a judgement that binds in the client's financial backers. Such a judgement may be based on information supplied by the interested firm itself, rather than an independent source. Thirdly, that men and women of integrity do exist who will stand by their considered views, even when the going gets rough. Fourthly, that politicians are capable of overruling the soundest economic advice (HoCFAC 1994: vol. II, 289–93).

5.10 Final remarks

The application of social cost–benefit analysis in the water sector constitutes part of the much broader history of project evaluation in recent decades. Little & Mirrlees (1991) suggest that from the late 1960s there was a considerable development of the method in respect of developing countries, with widespread application beginning in the 1970s. The use of shadow pricing was seen as particularly appropriate in economies where the allocation of resources was distorted primarily because of government taxes and controls. The World Bank adopted SCBA with a sophisticated use of shadow prices in a major way, as did some national governments, such as that of India, as well as bilateral aid agencies such as the British ODA.

However, the degree of penetration of project evaluation in this form has been limited, for public projects derive from government ministries or other agencies with departmental, sectoral or regional interests. SCBA using a standard methodology is often seen as the deployment of the power of central government's ministry of finance or ministry of planning, and has been resisted by other departmental rivals.

Within the World Bank, the use of a full range of shadow prices is seen to have declined in the 1980s and project evaluation in general has been reduced with the relative decline in project lending. Structural adjustment and other non-project loans account for a rising proportion of World Bank lending.

Little & Mirrlees argue that the use of SCBA for public sector investments is non-existent in most developing countries and in only a few is it better than rudimentary. Moreover, ". . . in only a minority of developing countries is there any coherent and rational control over ministries and other public agencies, with the result that far more projects get started than can be financed without interruption" (1991: 372).

The potential applications of project evaluation in the field of water are wide, in the form of either SCBA or SCEA. They embrace both river basin functions and water utility services, and can handle both supply augmentation and demand management.

This chapter has outlined in a brief space the principal characteristics of social cost–benefit analysis. Two adaptations of the simple model seem to me to be particularly important in respect of water infrastructure projects.

The first adaptation is not so much to the basic model but to how it is practised. It is argued here that net present value evaluation is the right approach, rather than its close cousin, an internal rate of return cut-off. Where projects are fiscally interdependent, the net benefit–investment ratio should be used. The use of the NPV/ NBIR technique pushes policy-makers into considering what the appropriate social time rate of discount (STRD) is for their country, whereas the use of the IRR introduces an unjustified bias against projects with higher (but back-loaded) undiscounted returns, whenever the IRR cut-off, typically 10 per cent, exceeds the appropriate STRD. Moreover, the NBIR substitutes for the IRR's opportunity cost of capital function.

The second adaptation is to use shadow prices rather than market prices, principally in the case where market prices of inputs do not reflect their opportunity costs. Shadow pricing will strengthen the relative NPV of water infrastructure projects, insofar as they use unskilled labour and locally manufactured construction materials. But shadow pricing weakens such projects' relative NPV insofar as they produce no exports nor import substitutes, and they rely heavily on, for example, imported monitoring and control equipment, imported plant and overseas consultancies.

The chapter contains two case studies. In Peru, the focus was SCBA and SCEA applied to the fresh- and wastewater utility of a medium-size town where short-term projects were drawn up in 1995–6 to rectify base-period weaknesses of an institutional, commercial and operational character. These projects required restructuring finance, which would be provided, indirectly, by the Inter-American Development Bank. IADB's SIMOP model was used for SCBA purposes. The model is derived from the neoclassical paradigm. Its weaknesses are the inconsistent employment of the concept of use value; the assumption of linear or, alternatively, constant-elasticity demand functions; the assumption of a zero-elasticity supply function; an inadmissible use of the concept of consumer surplus; and the exclusion of overhead costs from price determination. Real economic relationships are falsely represented, in order to create a working simulation model, rather than the model being designed to represent reality. In this instance, the view that neoclassical economics is "a methodology which sanctifies fictions and diverts attention from the difficult task of analysing real world phenomena" is confirmed (EAEPE 1991a).

In Malaysia, the decision-making with respect to the Pergau hydroelectric power project illustrates the use of cost-effectiveness analysis in the timing and sequencing of technically interdependent projects. It also shows that in aid programmes where proposals arise from bilateral negotiations between client and contractor, and where the award of a contract is a source of major exports from the aid-giving country, the process of economic appraisal may be exposed to intense commercial and governmental pressures.

CHAPTER SIX
Financial accounting for water enterprises

6.1 Introduction

In Sofia (Bulgaria) in September 1991, an environmental programme for the Danube river basin was conceived, with the goal of improving the basin's environmental management. The Danube is 2857 km long and its basin covers 817 000 km² in 17 countries in the heart of central Europe. By 1995, the programme's task force had produced its strategic action plan for the Danube river basin 1995–2005 (EPDRB task force 1995). The document's third chapter, "Financing plan implementation", reviews both the resources needed for the programme and likely sources of funds. In respect of loans from international financial institutions, emphasis is placed on the capacity of the borrower to repay the loan, and the Task Force writes:

> Maximising the involvement of the private sector takes the burden off central government and effectively implements the polluter pays principle. This approach, working closely with the enterprise or company to develop the project from the bottom up, is currently being successfully developed in Slovenia, Hungary and Romania by EBRD.

This link between the project and the private sector enterprise, within a context stressing the financial ability of the borrower to repay a loan, makes a strong bridge between Chapters 5 and 6. The focus of Chapter 5 was the project, benefits and costs from a social point of view, and the language of the economist. In contrast, Chapter 6's orientation is to the enterprise, financial flows from the private point of view, and the language of the accountant.

Nelson (1994: 242) said that ". . . the actions of individuals and organizations in society are largely determined by the socially learned and accepted pattern of behaviour, the routines, they follow." Here we turn from the analytical routine of social cost–benefit appraisal to that of financial accounting.

6.2 Balancing the books

The day-to-day practice of accounting varies between nation-states and between types of institution. However, the principles of accounting that were first developed in Europe in the twelfth century are now widespread throughout the world in the modern manufacturing, utilities and services sectors. In this chapter it is assumed that one is dealing with a **public limited company** (plc), Alzbeta Bathori Fresh- and Wastewater plc, incorporated in the UK with the motto "blood is thicker than water". A public limited company is a profit-seeking private company, the shares in which are available for sale to the general public and in which the owners' liability, should the company be bankrupted, is limited to the loss of their share capital. However, it is worth noting that water utilities, across the world, are often not-for-profit entities, although the process of utility privatization is changing that.

J. R. Dyson (1994: 12)[1] defines financial accounting as the preparation of highly summarized financial information, usually presented for the owners of an entity but also used by management for planning and control purposes, and of interest to the public, government, employees, investors, lenders, creditors and consultants. Here, the focus is on just three types of tabulation: the **profit and loss account**, the **balance sheet** and **cashflow statements**.

As a first step, one should understand that the financial affairs of an enterprise are recorded in a great variety of separate accounts. These might include the company's bank account; a creditors account recording what the entity owes its suppliers for goods and services supplied to it on credit; a debtors account recording what is owed to the entity by its customers for goods and services sold to them on credit; a cash account for notes, coins and cheques received prior to their transfer to the bank; a sales account recording the value of goods sold to customers during a particular accounting period; and so on and so forth. The main time-period used for financial accounting statements is one year.

Within the full set of a company's accounts, every transaction is recorded using double-entry book-keeping; that is, each transaction is recorded twice. In one account (the account that provides funds for the transaction) the transaction is recorded as a credit item. In a second account (the one that receives in the transaction), the same transaction is noted as a debit item. For example, the transaction of paying cash into the bank will be recorded as a credit in the cash account and as a debit in the bank account; the transaction of a cash receipt for the sale of goods is recorded as a debit item in the cash account and as a credit item in the sales account. So, any single account has two sets of entries in the ledger book: debit items are recorded on the left-hand side and credit items are recorded on the right-hand side. In many companies, the traditional ledger book has been replaced by computer accounts programs or spreadsheets.

For the enterprise as a whole, another way of determining the treatment of individual transactions is by using the fundamental accounting identity:

---

1. A rewarding introductory text on accounting, from which this discussion is derived.

$$A - L \equiv E \qquad\qquad (6.1)$$

where A is assets, L is liabilities and E is equity. Where a transaction adds to assets, it is recorded as a debit item – such as in the case of the cash purchase of a machine. Where assets are reduced by a transaction, it is recorded as a credit item. The converse rules apply to an increase or fall in liabilities. We return to the concepts of assets, liabilities and equity in §6.5.

During any accounting period, a single account may show a large number of debit and credit entries. Balancing an account requires all the debit and credit items to be separately totalled. It will be rare for the two sums to be equal. Where debits exceed credits, a debit balance is said to exist and, to make the two sides of the ledger balance, is entered on the credit side. A similar procedure takes place where a credit balance exists, that is, where the account's outgoing items exceed the value of its incoming items. Here, the balancing item is entered on the debit side.

This leads us to the trial balance. As J. R. Dyson writes (1994: 55):

A trial balance is a statement compiled at the end of a specific accounting period. It lists all the ledger account debit balances and all the ledger account credit balances. A trial balance is a convenient method of checking that all the transactions and all the balances have been entered correctly in the ledger accounts. Once all the debit balances and credit balances have been listed in the trial balance, the total of all the debit balances is then compared with the total of all the credit balances. If the two totals agree, we can be reasonably confident that the book-keeping procedures have been carried out accurately.

After the trial balance stage, and prior to presenting the basic financial statements, four main types of final adjustment are required: for stock, depreciation, accruals and prepayments, and bad and doubtful debts. Stock adjustment reflects the difference between the value of the opening stock of a firm, at the beginning of the accounting year, and the closing stock. By their very nature, fresh- and wastewater utilities have no stock valuation of their outputs, although changes in the value of input stocks, such as chemicals, cannot be disregarded. Adjustments for depreciation, accruals and prepayments, and bad and doubtful debts, are discussed in the next section.

6.3 The profit and loss account

The profit and loss account of a private company is, as the name suggests, a financial statement containing entries that show for an accounting year whether the company made a profit from its activities during that year, or a loss. To prepare it, all the company's accounts must first be balanced and, next, capital accounts must be separated out from current accounts. The distinction between the terms "capital" and "current" has already been discussed in §3.1. Accountants use the same concepts as economists but prefer the terms "capital" accounts and "revenue" accounts. Here we stick with the economists' terminology. Profit (or loss) is defined as the difference between current income and current expenditure.

At this point, it is essential to understand that the accounts which form the input to the profit and loss calculation are based on the matching rule. For any given financial year, 1995/6 for example, all expenditure and income should relate to the economic activity of that year, even where an input such as chlorine is not paid for by the company until 1996/7 or where the supply of fresh water is not paid for by customers until 1996/7. Accountants adjust cash paid and cash received on what is known as an accruals and prepayments basis to allow for this lack of synchronization between economic activity and the cashflows associated with it.

Since the profit and loss account is derived only from the current accounts of the enterprise, it follows that capital expenditures such as the construction of a reservoir, the installation of switchgear and the laying of sewage pipes are not directly represented in the profit and loss calculation. However, these fixed assets are certainly necessary to generate sales income and therefore the cost of this benefit needs to be represented in some way if profit is not to be over-estimated. This is done by setting a charge for depreciation. This corresponds to the references to amortization in §3.3 and Figure 3.2.

More than one method exists for the calculation of depreciation. The most common is the straight-line technique. Here one estimates the working life of the fixed asset, 20 years for example, dividing this into the historic cost of the reservoir, switchgear or piping, and the resultant figure is used as the depreciation charge throughout the asset's life. Note that the standard practice of accounting is to use historic cost entries. However, with some major and long-lasting items, the depreciated value of the asset may be revalued upwards periodically to reflect inflationary movements in new asset prices. Revaluation leads to a higher recorded depreciation charge.

Another technique for generating the depreciation entry is the reducing balance method. In each year the charge is set equal to a given percentage of the asset's value at the beginning of the year. That charge is deducted from the asset value to give a recalculated value for the start of the following year. The effect of this is to record much higher depreciation charges in the asset's early life in comparison with the straight-line approach.

Alongside stock changes, accruals and prepayments, and depreciation, the fourth post-trial balance adjustment referred to in §6.2 is in respect of bad and doubtful debts. As we have seen, the matching rule requires a company to record as sales all goods transferred to its customers in a given year, even though payment for them may not have been received in that year. However, experience teaches that some of these customers never pay. In the financial year when a debt is recognized to be irrecoverable, it is referred to as a bad debt and is deducted from current income prior to profit calculation. Similarly, provision is made for the possibility of future bad debts. When these doubtful debt provisions, often set equal to a percentage of the trade debtors account, increase from one year to the next, that change is deducted from current income. Bad debts and doubtful debts were an important feature in water utility accounting in the transitional countries in the years following the disintegration of the Soviet Union.

Before the preparation of the profit and loss account takes place, it is common to derive two prior financial records from the full set of the company's double-entry accounts. One sets out the income of the company. In the case of Bathori plc, this is derived primarily from the sale of freshwater supplies to its customers and from the charges it levies for wastewater treatment. A second account records the cost of operations. This will be derived from the firm's cost- and management accounting practices. We can call these the trading and processing accounts.

The processing account includes the cost of materials and spares, wages and salaries, power supplies, plant and equipment depreciation and miscellaneous items such as carriage inwards for materials. (Materials costs will be set equal to the value of the opening stock of materials plus material purchases minus the value of the closing stock.) Where these five items are clearly capable of allocation, either to freshwater supply or wastewater treatment, this is done. For costs that cannot be so allocated, such as company administration, rent and rates on premises, and heating and lighting, they may be brought together in indirect cost categories or, in spite of the difficulties, allocated between the two main product groups on some rule-of-thumb basis. The attraction of allocating all current costs in this way is that the cost of fresh water is distinguished from that of wastewater and this can form the basis of product pricing, as was shown in §4.6.

The costs of the processing account are subtracted from the income of the trading account to give gross profit. We can now turn to the profit and loss account. The first item is gross profit, as calculated above. From this may be deducted some office, selling and distribution expenses not already included in the two prior financial statements. Discounts received will be added and discounts allowed subtracted. Finally, loan interest is subtracted to give net profit. Note that the treatment of gross profit in §3.3, as sales less prime costs, is not the same as the accounting definition here. For example, the accountant subtracts some components of depreciation prior to the gross profit calculation. However, net profit is the same in both approaches.

Finally, Alzbeta Bathori plc's profit and loss appropriation account shows what happens to profits. The account begins with the firm's net profit for the financial year. From this is subtracted first the corporation tax payable to the Inland Revenue, and then the dividends payable to the company's shareholders. What is left is the retained profit for the year, which, when added to the retained profits brought forward, gives the retained profits carried forward to the next financial year.

6.4 Command over resources

The next financial statement to be considered here is the company's balance sheet. But before doing that, a more general discussion is required of the sources of finance available to a public limited company. As Robinson & Eatwell write (1974: 99):

> A financial mechanism that provides firms with a command over resources beyond what they own is a basic characteristic of a modern capitalistic economy. It derives from the ability of capitalists to inspire confidence of future gain in those who lend to them. The less confidence they inspire, the higher the premium they must pay on borrowed funds. But if capitalists as a whole ceased to inspire any confidence at all, this would entail a collapse of the financial system, and indeed, of the whole economic system.

A firm's "command over resources" derives from six distinct sources, to be reviewed in turn. These are share capital, **debentures**, loans, **overdrafts**, **trade credit** and retained profits. This last item is addressed in §6.5.

When a plc is first created, its capital derives from the purchase of the *shares* that it issues. Ownership of the company exists through the ownership of these shares. The maximum sum the company envisages raising, at least at its inception, is called its authorized share capital. This, in the UK in 1995, had to be at least £50000. Shares could be issued in denominations of 10p, 50p and £1, for example, but the price actually paid for each share would depend on many market factors at the time the company is formed.

The amount of shares actually issued is known as the issued share capital. The two main types of share are ordinary shares and preference shares. The former do not entitle the shareholder to any specific level of dividend out of profits. Preference shares are normally entitled to a fixed level of dividend, but only if the company makes profits, and they have priority of repayment over ordinary shares if the company is liquidated (J. R. Dyson 1994: 134–40). Dividend payments to shareholders have already been referred to in discussion of the profit and loss appropriation account.

Debentures are legally a specialized loan with a trustee to look after the holders' interests. They are an acknowledgement of debt and they carry a promise to pay a certain sum of interest each year. They carry no rights of ownership in the firm. The interest payment is a legal obligation. The debenture's redemption date is when the original sum loaned must be repaid. Both shares and debentures are bought and sold on the Stock Exchange. As we have already seen, interest (including debenture interest) is chargeable as a business expense against gross profit in calculating the net profit on which corporation tax is levied. This is not true of dividends.

A *loan* is a transfer of money from a bank, for example to an enterprise. The accompanying legal documentation will specify how the principal of the debt is to be repaid, as well as the rate of interest payable on the outstanding debt at any time.

The outstanding debt is the original sum lent less the principal repaid. To take an actual case, in 1995 the World Bank set out an example of its terms for a $1 million loan to a fresh- and wastewater enterprise. The accounting periods were six-monthly, with semi-annual payments on 30 June and 31 December. The entire loan was assumed to be taken up over the first nine periods following the set-up period. A commitment fee of 0.25 per cent per period was required on the sum not taken up, payable annually on 31 December. The term of the loan was 17 years. There was a grace period of 5 years, so that repayments of the principal of the loan did not begin until year 6 and from then on was in equal absolute amounts per period. The interest rate was 7 per cent, payable from the end of the first period in which the loan was disbursed. On a loan of $1 million, the total payments over the 17-year period at constant prices would be equal to $2 079 883, so for each dollar borrowed two must be paid back.

The loans provided by national export–import banks, or loans guaranteed by export credits agencies, are tied to a company's import of capital goods from the lending country. They are of great importance in build–own–operate–transfer (BOOT) ventures, such as those in Scotland and Turkey (Merrett 1997)

The fourth and fifth forms of command over resources referred to above are *overdrafts* and *trade credit*. These can be dealt with briefly. An overdraft is an agreed limit by which, from the bank's point of view, the company's bank account can be in debit. A company pays a rate of interest on the actual sum overdrawn. Trade credit is the period of grace conceded by materials and equipment suppliers between the date their goods are delivered and the date by which payment is required for them. Trade credit can be considered as a short-term, interest-free loan.

6.5 The balance sheet

Now one returns to the balance sheet. In describing the nature of the profit and loss account, it was pointed out that, at the trial balance stage, the totality of an enter-prise's accounts for any year should be divided into its capital and current accounts. The profit and loss account is derived from the current balances. The balance sheet, in contrast, is derived from the capital accounts and can be regarded as a summary statement of their position at the *end* of the financial year. (The profit and loss account is a statement covering the *whole* year.) In effect, the balance sheet sets out the enterprise's assets (what it owns or is owed) and its liabilities (what it owes). The excess of assets over liabilities is defined as equity, or total shareholders funds, as has already been set out above in the fundamental accounting identity.

The company's assets are divided into fixed assets and current assets. Fixed assets refer to shares held in other companies, other long-term financial invest-ments, land, buildings, plant and equipment, and vehicles. Against each item or class of real assets will be set its historic cost (gross book value), the accumulated depreciation on that item or class, and the excess of the former over the latter, that is, the net book value. Gross book value may be based on a revaluation of the assets to reflect specific price changes in land, new equipment, and so forth. Current assets will include stocks of raw materials, work in progress and finished goods, trade debtors less provision for doubtful debts, bank deposits, cash and prepayments. This is the firm's working capital.

The firm's current liabilities include trade creditors, accruals, its bank overdraft, corporation tax due for payment after the end of the financial year, and proposed dividends not yet paid.

Fixed plus current assets less current liabilities are then aggregated. The net total of assets over current liabilities is then expressed in terms of how it has been financed. Such financing is divided into two categories. The *first* is made up of share capital (ordinary and preference shares issued and fully paid up) and reserves (retained profits and other reserves). Retained profits are a most important source of funds for capital investment. Capital reserves are sums not available for distribution to shareholders as dividends. The *second* financing category is loans, composed of debenture stock and other long-term loans. The items of which a balance sheet is composed is set out again in Table 6.1.

6.6 Cashflow statements

Traditionally the profit and loss account and the balance sheet have been the most important financial accounting statements available to an enterprise's board of directors. However, in recent years, a third document, the cashflow statement, has also come to be seen as providing a distinct and valuable insight into the financial affairs of a company. The reason for this is that neither the profit and loss account nor the balance sheet give an adequate picture of the firm's cash position. Cash is here understood to mean cash in hand in notes and coin, deposits in a bank or other financial institution repayable on demand, investments readily convertible into known amounts of cash and with a maturity of less than three months, less bank advances repayable within three months.

The critical importance of the cash position is that a shortage of liquid funds prevents an enterprise paying its creditors and employees. In the UK, this would make it insolvent, that is, the firm could not lawfully carry on in business. The outcome could be the break-up of the company or takeover from a rival. The attraction of the cashflow statement is precisely that it describes in detail the movement in an entity's cash position during a defined period.

A cashflow statement can be prepared in more than one way. The approach here is known as the indirect method (J. R. Dyson 1994: ch. 7). The statement's basic structure is composed of operating activities, investments and the servicing of finance, taxation, investing activities, financing, and the net total, the increase/decrease in cash.

Table 6.1   Balance sheet items.

Item	Value
Fixed assets	
Shares in other companies	
Other long-term financial investments	
Land	
Buildings	
Plant and equipment	
Vehicles	
Other fixed assets	
Total fixed assets	A
Current assets	
Stocks: raw materials	
Stocks: work in progress	
Stocks: finished goods	
Trade debtors (Less provision for doubtful debts)	
Bank deposits	
Cash	
Prepayments	
Total current assets	B
Current liabilities	
Trade creditors	
Accruals	
Overdraft	
Corporation tax due for payment	
Proposed dividend	
Total current liabilities	C
Total net assets	A+B–C
Financed by:	
Capital and reserves	
Capital: ordinary shares	
Capital: preference shares	
Reserves: accumulated retained profits	
Reserves: capital reserves	
Total shareholders funds	D
Loans	
Debentures	
Other long-term loans	
Total loans	E
Financing total	D+E

N.B.: A+B–C must equal D+E.
Also: A+B–C–E = D = Equity, that is, total shareholders funds.

Operating activities are taken from the profit and loss account after a series of adjustments to convert operating profit, before taxation and before dividend payments, into cashflow. This is done, in some cases, by adding to recorded profit. Such additions include the depreciation charge, bad debts written off, decreases in stocks, in trade debtors and in prepayments, increases in trade creditors and in accruals. For example, depreciation is added in to calculate cashflow because the original deduction to calculate profit did not reflect any cash outflow for the firm. In other cases, conversion from operating profit to cashflow requires deductions and these include increases in stocks, in debtors and in prepayments and decreases in creditors and in accruals. For example, an increase in debtors implies less cash coming in to pay for water bills.

Investments and the servicing of finance includes dividend and interest payments as negative entries, and interest received by the firm as a positive entry. Taxation refers to all taxes paid during the financial year, of which corporation tax is the most significant. Investing activities includes as a positive entry cash received from sales of fixed assets and, as a negative sum, cash paid for the purchase of intangible and tangible fixed assets. Financing activities include cash paid on the redemption of debentures and cash received from the issue of shares and debentures. The net total must be reconciled with the opening and closing cash balance shown in the balance sheet.

6.7 Ethical principles and consultancy practice

Chapter 5's discussion of social cost–benefit analysis pointed out that project approval, in almost every case, will be in the strong material interest of some social or economic groups, but will disadvantage others. As a result, insistent pressure may be brought to bear on the economist to arrive at a final appraisal favouring one group rather than the other. This pressure, insidious or brutal, seeks a dishonest manipulation of the SCBA technique or its primary data and challenges the integrity of the analyst.

This *leverage for bias* also occurs in financial accounting. Dyson notes the subjective nature of accounting practices with respect to the end-of-year valuation of unsold goods, to the distinction between capital and current items, to allowances for accruals and prepayments, to depreciation estimation, to the treatment of doubtful debts and to the adjustments to be made for inflation once one leaves the tranquil waters of historical cost accounting. These areas all offer potential for "cooking the books", an old and revealing metaphor.

Within the accountancy profession, to its credit, certain ethical conventions have evolved to counter bias. The consistency convention requires that, once specific accounting practices are adopted, they should be maintained in subsequent financial years. The objectivity convention states that accounting should be carried out with the minimum of bias. The relevance convention requires the selection of materials for inclusion in financial statements, from the mass of information that is available, to be done so that "a true and fair view" of the company's financial status is given. There is also a prudence convention which states that, where there is room for doubt, losses should be overstated and profits understated. However, this last convention itself seems a form of bias (J. R. Dyson 1994: 30–34, 93–5).

In the water industry it is evident that the structural relation between the client and a consulting firm forms the basis of leverage for bias. When, for example, an engineering and economic feasibility study is required, the client – such as a ministry of the environment – will first have the terms of reference prepared. The competitive bidding process between firms then takes place and the winner is selected, hopefully on the quality and price of its technical proposal. The work begins and a steering committee is set up to follow progress through to the draft and final reporting stages. The consulting firm's project team leader is a key player in all this. This person is often a victim of excessive workload and consequently is incapable of developing effective team relationships.

The consultancy is always in search of fresh work and the client may well be a source of future contracts. So, the firm is eager to please. If, because of the power relations within the ministry, the client has a strong view on the desired outcomes of the study and presses its case, the company may face a dilemma. The objective view of the team may indicate a direction inconsistent with the client's wishes. However, few consultants will clearly oppose the client's line. So, either the final report fudges the key issues or sets up the analysis to deliver whatever the client seeks. The client then represents the final report as the considered view of "an independent consultant".

This corrupting nexus can also exist when there is what one might call a "shadow client". The formal client may have inadequate human resources to establish its own terms of reference, to adjudicate between rival technical proposals and to steer the study's progress. This can open a space for another body to take over the client's role. The World Bank, for example, may assume the mantle of the shadow client.

The Bank is, of course, a most powerful institution. Its financial resources are immense, its staff are of great technical competence, it controls an unrivalled information base, and it has a well developed philosophy of action. In one case, a ministry had no funds and the World Bank looked to be the only potential lender to a water utility in a town desperately in need of fresh investment. It was the Bank, above all, which became the source of guidance on the day-to-day work of the consulting team. Indeed, the Bank – and not the client – insisted on the rejection of a technical appendix to the draft report with which it disagreed, and the consulting firm began to speak of the Bank as the client, not the ministry. As one team leader said: "Remember, the World Bank is always right."

In another case, consultants modelled a profit and loss account, balance sheet and cashflow statement for a local utility. The Bank informed the team that the critical tabulation would be of the cashflow, and that there should be no negative cash balances. Using the team's tariff-setting criteria, first runs of the model revealed negative cashflows in all the options under investigation. The response was to increase the tariff above the stated tariff criteria to eliminate the negative flows, and to present those results to the client.

6.8 The financial analysis of projects

At this juncture, it is necessary to point out a gap in the analytical scope of Chapters 5 and 6. Chapter 5 dealt with the appraisal of projects from a social perspective. Chapter 6 deals with financial accounting for the firm from a private perspective. What is missing is the private appraisal of projects, and that is the subject of this section. As in the case of Chapter 5, it is assumed that there are alternative projects, each with a flow of differential real outputs and real costs in comparison with the no-project scenario, and that these flows can be estimated in market price terms, giving a net benefit stream of the type illustrated in Table 5.1. It is also assumed that the enterprise faces financial or managerial constraints on its investment in any year so that not all technically independent options can be selected.

Five forms of appraisal are reviewed below. All these options are used in practice, although the final decision to launch a project will as much reflect the "animal spirits" of the senior management as any arithmetic tabulation. These five forms are as follows: rate of profit on capital, payback period, net present value, internal rate of return, and the discounted profit:capital ratio. In each case, the precise formulation will depend on the definition of such concepts as profit, capital and interest rate.

Rate of profit on capital
In the accountancy world, the rate of profit on capital technique is also known as the accounting rate of return. Capital investments are carried out in the search for profits, and so appraisal in terms of a profit:capital ratio, expressed as a percentage, is an obvious choice. Net annual profit can be calculated and divided by the total capital outlays required to generate it. The drawback to this approach is that it does not reflect the time-pattern of outlays and income. The social time rate of discount, discussed in Chapter 5, has no part in private decision-making, but some form of discounting is appropriate. This is because capital expenditure generates time-dependent costs: if funded from debt, interest payments must be made; if paid for from retained profits, other interest-bearing opportunities are forgone. For a given time-stream of profits, the later the capital outlays on a project take place, the lower their interest costs. Similarly, the earlier the flow of profits, for a given capital outlay, the sooner the debt can be repaid.

Payback period
The payback period is measured by the ratio of total capital expenditure to annual net cashflow after interest payments and taxation. It can be thought of as the number of years that would elapse before net cashflow enables the capital sum expended to be paid off. More sensitive to the timing of profits, its weakness is that it completely ignores costs and returns after the payback period is over.

Net present value

Here, the net benefit stream is discounted to its present value. The discount rate could either be the rate of interest payable on loans at the margin, or the rate of interest forgone on retained profits, or a weighted average of the two, reflecting the financing balance between external and internal funds. Fully sensitive to the timing of costs and returns, this approach has the fatal weakness of not reflecting how project size influences net present value and is therefore unsuitable where constraints exist on total investment, imposing choices between technically independent projects.

Internal rate of return

As in Chapter 5, the internal rate of return is the discounting rate which, when applied to the stream of net benefits, gives a value of zero for the discounted total. Fully time-sensitive, it can also be used to rank investments. In spite of some serious criticisms in the technical literature, it is widely used.

Discounted profit:capital ratio

Here the numerator is the discounted net benefit stream during the project's working life and the denominator is the discounted capital cost of the project during its gestation period. This approach corresponds to the net benefit–cost ratio of social cost–benefit analysis. This has the same positive features as the internal rate of return and is more easily comprehended.

Whatever the approach adopted in project selection, the risk–return principle is certain to be applied, that is, the higher the perceived risk of an investment, the greater will be the return required from it. Project risk is greater, the higher the variance of the estimated benefit and cost outcomes, and the more severe is the inflexibility of the commitment (Robinson & Eatwell 1974: 178–9).

Investment by a public limited company in a fresh- and wastewater utility is, of course, highly inflexible. If a project goes badly wrong, little of substantial value could be retrieved. The degree of variance of costs and returns is more complex. The absolute dependence of human populations on water for life and the natural monopoly of a utility reduces the variance in benefits. However, as was illustrated in the Latvian case study of Chapter 4, the sale of water to households, industry and farmers is critically affected by the health of the local economy. In respect of costs, the role of government in regulating the quality of drinking water supplies and of wastewater treatment can impose unexpected increases. Similarly, the setting of abstraction and discharge fees by government, its regulation of the price of water and its policy towards demand management – such as through domestic metering – will be of major concern to the industry. For these reasons, the private water sector is certain to seek a strong position of influence on government policy, through the efforts of individual companies and through trade associations.

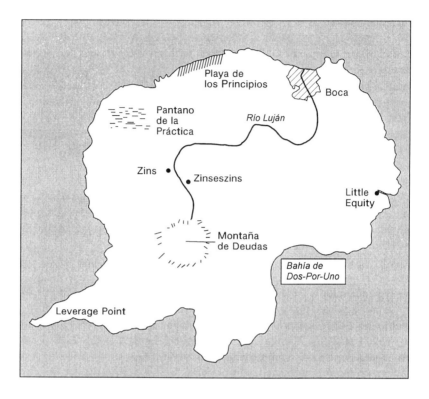

Figure 6.1   San Serif.

6.9 Case study: a fresh- and wastewater plant in San Serif[2]

San Serif is a large island in the area known as the Bermuda Triangle (Fig. 6.1). Its capital is Boca, with a population of some 100 000 people, located at the mouth of the Río Luján. The river is seriously polluted because of upstream activities and because much untreated wastewater is discharged into the river in the capital itself. The local fresh- and wastewater utility – San Serif Water – dates from the colonial period in the nineteenth century. Owned by the municipality, it is antiquated and it provides neither good quality drinking water nor adequate treatment of the city's sewage. San Serif's currency is the peso, divided into 100 centavos. The rate of exchange of the peso to the dollar is one to one.

The United Nations regional commission, at the request of San Serif's Ministry for Environmental Protection, has commissioned a feasibility study into the supply of fresh- and wastewater services in the capital. Bathori Consulting, a wholly

2. In this case study, names have been changed to preserve commercial confidentiality.

owned subsidiary of Alzbeta Bathori plc, has secured the contract for this work. The professional grapevine has it that Bathori Consulting have been successful against much-fancied competition because, unlike the other engineering consultancies, the parent company is itself a water utility. There are real possibilities that San Serif Water will be privatized and, informally, Alzbeta Bathori plc have indicated to the Prime Minister that it would consider carrying this through.

The financial tabulations at the heart of Bathori Consulting's study are set out in Tables 6.2–6.5. In fact they are all drawn from a single Excel™ worksheet 25 columns wide by 550 rows deep. The bulk of the space in this worksheet is taken up by a series of subroutines to which cells in the main published table cross-refer. For example, a subroutine exists for the time-stream of potential loan disbursements from the World Bank, alongside the consequential payments of interest, repayments of principal, and commitment fees. A cell indicating the sum of interest plus fees in the main table simply picks up the output in a defined cell of the subroutine. Other subroutines include tariff calculations, two sets of assumptions on Boca's average household income growth, and depreciation schedules.

Another point of clarification is necessary. Tables 6.2–6.5 are the results for just one option out of four considered: there are two orientated to improvements in the freshwater supply and two for investment in wastewater treatment. Curiously, no option for capital expenditure in respect of both fresh- and wastewater is considered. All four tables here consolidate the separate results and tabulations for San Serif's freshwater and sewerage functions.

Table 6.2 provides vital information on the size of the local market and the level of tariffs. The full projections are made through to the year 2010, but that level of detail serves no useful purpose in this case study, so the tabulations here are limited to 1993–7. There are several points of interest. Total population is falling as a result of a severe, long-term industrial recession and also because of serious (and often violent) divisions between the island's two main ethnic groups. No parallel decline in the number of industrial connections is forecast, but that is a case of the consultants whistling in the dark, for no substantial research of any kind was carried out during the demand forecasting exercise. Domestic consumption is shown to the third decimal place, but in fact no-one knows how much water households consume since no domestic meters exist. Similarly, the true ratio of unaccounted-for-water to total production is unknown, and no local expert believes that the 74 per cent decline in unaccounted-for-water in the two years after 1994 is feasible. Finally, the large variation in tariffs between the three sets of customers disappears from 1996 onwards because, from that date, it is assumed all tariffs will be set on the same basis.

The statement of income and expenditure in Table 6.3 combines accounts for sales, process costs, and profit and loss. The most interesting feature under billed water sales and sewerage charges is the leap in income from domestic households after 1995. This is the result of a doubling in 1996 of the domestic tariffs for water and sewerage, as shown in Table 6.2. As a consequence, total operating revenue increases by 78 per cent in that year.

Table 6.2   San Serif Water: the market.

		Actual 1993	Actual 1994	Estimated 1995	Projected 1996	Projected 1997
Total population	000	117	110	107	105	103
Population served	000	94	86	86	86	86
Population served	%	80	78	80	82	83
Water supply connections						
Domestic	nos.	34020	34470	34457	34457	34094
Industrial	nos.	450	450	450	450	450
Municipal	nos.	270	270	270	270	270
Total	nos.	34740	35190	35177	35177	34814
Domestic consumption						
Av. per capita consumption	m^3 per day	0.284	0.284	0.28	0.276	0.272
Average household size	persons	2.75	2.48	2.48	2.48	2.48
Water production						
Billed water sales:						
Domestic	m^3 million	7.74	6.93	6.39	7.8	7.72
Industrial	m^3 million	1.26	1.44	1.71	1.74	1.76
Municipal	m^3 million	1.53	0.99	0.81	0.81	0.81
Total	m^3 million	10.53	9.36	8.91	10.35	10.29
Unaccounted for water	m^3 million	3.15	3.15	1.8	0.91	0.91
UFW % of production	%	30	34	20	9	9
Water tariffs: outturn prices						
Domestic	pesos	0.054	0.076	0.08	0.16	0.18
Industrial	pesos	0.1	0.18	0.18	0.16	0.18
Municipal	pesos	0.132	0.082	0.08	0.16	0.18
Sewerage volumes						
Domestic	m^3 million	8.01	7.02	6.3	6.75	6.75
Industrial	m^3 million	3.51	3.51	3.51	3.78	3.78
Municipal	m^3 million	1.53	1.17	1.17	1.35	1.26
Total	m^3 million	13.05	11.7	10.98	11.88	11.79
Sewerage tariffs: outturn prices						
Domestic	pesos	0.028	0.058	0.06	0.12	0.12
Industrial	pesos	0.088	0.108	0.132	0.12	0.12
Municipal	pesos	0.06	0.06	0.06	0.12	0.12

With respect to operating expenses, one notes in 1995, for example, that, although personnel makes up 29 per cent of the total, all nine categories make a significant contribution to costs – there is no question of just two or three categories dominating the account. Personnel costs fall after 1995 because plans exist to "downsize" San Serif Water's workforce by 12.5 per cent. Chemicals rise sharply after 1995 because access to subsidized chlorine is assumed to terminate. The leap in depreciation in 1996 reflects the option that Tables 6.2–6.5 seek to appraise: a major investment in Boca's sewerage system. The shift in bad debts from zero in 1994 to 160000–182000 pesos in 1995–7 reflects the fact that for several years San Serif Water has faced a problem of unpaid bills with its domestic, industrial and

Table 6.3 San Serif: Statement of Income and Expenditure in '000 pesos.

	Actual 1993	Actual 1994	Estimated 1995	Projected 1996	Projected 1997
Consolidated operating revenue					
Billed water sales					
Domestic	360	500	508	1280	1408
Industrial	130	266	311	284	320
Municipal	200	83	68	133	148
Other	7	22	14	0	0
Subtotal water	697	871	901	1697	1876
Billed sewerage charges					
Domestic	176	378	374	869	806
Industrial	306	376	459	488	454
Municipal	94	72	68	167	155
Other	5	14	7	0	0
Subtotal sewerage	581	840	908	1524	1415
Total operating revenue	1278	1711	1809	3221	3291
Operating expenses					
Personnel	414	619	653	540	544
Chemicals	76	220	292	391	450
Power	196	194	214	274	288
Environmental fees	293	229	304	344	360
Social tax	155	230	243	202	202
Heat	68	83	92	101	106
Depreciation	148	157	158	464	598
Other	−11	92	101	103	108
Bad debts	0	0	182	160	164
Total operating expenses	1339	1824	2239	2579	2820
Administration expenses	58	65	74	83	86
Operating income/deficit	−119	−178	−504	559	385
Non-operating income	29	65	54	54	54
Interest received	2	2	2	2	2
Grant received	0	85	0	0	0
Interest and commitment fees	0	0	0	0	0
Net income/deficit	−88	−26	−448	615	441
Turnover tax	0	52	103	146	140
Net income/deficit after taxes	−88	−78	−551	469	301

municipal customers. Households and business have suffered severely in the recession, and municipal customers cannot pay because of the decline in the local tax base. In 1995, unpaid invoices are equal to four months' operating revenue. San Serif Water had not previously recorded bad or doubtful debts, so here Bathori Consulting is confronting the situation head-on.

Operating income has been negative in 1993–5. The shift into the black in 1996–7 is primarily because of the planned tariff increases. What is of real concern to third party observers is that the tariff hike of 1996–7 in Table 6.2 is set alongside assumptions of significant increases in the volume of billed water sales and sewerage volumes in the same table. In its internal papers Bathori Consulting has never explained this perverse demand elasticity. No doubt this is because the technical paper on the consumption of water was written entirely independently of the tariff paper.

As for the profit and loss account, the only interesting feature is that there are no entries for interest payments and commitment fees. This is because Bathori Consulting, on the advice of the Ministry for Environmental Protection, has assumed that the sewerage investment will be 100 per cent grant-funded by a powerful neighbour in the region.

Table 6.4 sets out San Serif Water's balance sheet. Under fixed assets, one sees their valuation more than double in 1996 when donor funding is projected to be spent on upgrading the treatment plant. Under current assets, the cash position has deteriorated substantially in 1993–5, improving rapidly thereafter as the new tariffs take effect. Similarly, under current liabilities, the position improves greatly after 1995 as the backlog of overdue tax debts to Boca City are cleared. Long-term funding shows the receipt of the sewerage grants. Finally, total assets are equal to the sum of liabilities, equity and long-term funding.

Table 6.4   San Serif: balance sheet in '000 pesos.

	Actual 1993	Actual 1994	Estimated 1995	Projected 1996	Projected 1997
Fixed assets					
Cost/valuation	4541	5062	5078	12737	16067
Accumulated depreciation	2336	2493	2651	3116	3713
Net book value	2205	2569	2427	9621	12354
Current assets					
Inventory	47	110	126	319	401
Consumer receivables	476	818	842	1088	628
Other receivables	115	40	40	43	47
Cash	−32	−95	−231	−86	738
Total	606	873	777	1364	1814
Total assets	2811	3442	3204	10985	14168
Current liabilities					
Accounts payable	34	31	36	77	85
Taxes payable	365	529	767	441	59
Short-term loans	0	0	0	0	0
Other	41	0	0	0	0
Total	440	560	803	518	144
Equity					
Capital	155	155	155	155	155
Revaluation reserve	2239	2736	2736	2736	2736
Retained earnings	−23	−9	−490	−83	144
Total	2371	2882	2401	2808	3035
Long-term debts/grants					
Sewerage grants	0	0	0	7659	10989
Total	0	0	0	7659	10989
Total liabilities+equity+grants	2811	3442	3204	10985	14168

Table 6.5 is the cashflow statement. Net income after tax is drawn from Table 6.3, as is depreciation, which is added back in for cashflow purposes. The change in inventories shows a large cash commitment to restocking in 1996–7 and similarly for the change in taxes payable. Investment is offset by external financing in 1996–7 and debt payments are zero throughout the period. The end cash balance, identical to cash in Table 6.4, is positive only in 1997, showing the precarious financial position of the company. Had the new treatment plant been funded by a loan rather than a grant, then the enterprise could hardly have survived.

Table 6.5 San Serif: cash flow statement in '000 pesos.

	Actual 1993	Actual 1994	Estimated 1995	Projected 1996	Projected 1997
Beginning cash balance	7	−32	−95	−231	−86
Net income after tax	−88	−78	−551	469	301
Depreciation	148	157	158	464	598
Changes in working capital					
Accounts receivable	−218	−254	50	−306	385
Accounts payable	250	−4	5	41	7
Inventories	−34	−63	−16	−193	−83
Other receivables	0	76	0	−4	−2
Taxes payable	0	164	238	−326	−382
Other	−113	−41	0	0	0
Total	−115	−122	277	−788	−75
Investment					
Project	0	0	0	−7659	−3330
Other	−11	−20	−20	0	0
External financing					
Donor grants	0	0	0	7659	3330
Municipal grants	27	0	0	0	0
Debt repayments	0	0	0	0	0
End cash balance	−32	−95	−231	−86	738

What is not apparent in these tables, although it is made clear in the accompanying financial report, is that all the data are expressed in out-turn prices. Unfortunately, this makes it far more difficult to distinguish, in year-on-year calculations, real shifts from those merely attributable to inflation. In my judgement, such financial statements are best presented in constant prices, including relative price shifts as discussed in Chapter 5, and then recalculating the data for any inflationary movements the client wishes to anticipate.

Finally, it is worth noting in this case study that the financial accounting procedure combines a view of the enterprise as a whole with the presentation of results for four different options. This is done by setting out the profit and loss account, balance sheet and cashflow statement in four alternative versions, one for each of four mutually exclusive technical options. The drawback is that this set of financial accounts does not demonstrate which is the optimum choice in the project analysis terms set out in §6.8. This suggests that the terms of reference laid down for the consultants were primarily *enterprise*-orientated. A careful reading of these terms of reference confirms this, for they specify an interest in the development of an autonomous utility, that is, one which would be self-financing and politically independent of Boca City. From this one can surmise that the political agenda is for a future privatization – in which the overall financial status of the company is of far greater importance than project selection.

One can certainly ask why the financial report did not also include a traditional internal rate of return calculation for each project, since all the necessary data were available to do that. The answer is that, as a result of an intervention from a potential lender, Bathori Consulting rewrote the four technical options for evaluation the day before the report was submitted, and that the financial analysts received the new cost data from the engineers only eight hours before the deadline for handing in their report! In these conditions, the financial accounting for the enterprise was considered to be of much greater importance than the economic analysis of projects.

6.10 Case study: privatization of the water utilities in England and Wales

The government of Mrs Margaret Thatcher in the period 1979–90 must be counted as among the most radical in the UK during the twentieth century. At the heart of her declared philosophy was a reduction of the economic powers and responsibilities of government and the public sector, and a commitment to the self-interest of individuals and private institutions as the driving force for economic change. Mrs Thatcher declared there was no such thing as "society" and explicitly sought to create a "post-socialist" nation where the principles and practices of social democracy were but a memory of days long gone by.

The privatization of nationally owned assets was one means of delivering this philosophy, for it simultaneously reduced the scope of the public sector and extended that of the private sector. There were additional advantages of a fiscal nature, that is, with respect to government's getting and spending of money. Asset sales of this kind are treated by the UK Treasury as negative capital expenditure and therefore, in the short-term, reduce the level of the government's borrowing requirement. In the long term, fresh rounds of infrastructural investment no longer would be paid for by local authorities or the nationalized industries, but by the private sector. Kinnersley (1994: 4) – from whose splendid book *Coming clean: the politics of water and the environment* this case study repeatedly draws – suggests that, in the late 1980s, £24 billion was an agreed assessment of the required investment total in the English and Welsh water utilities over ten years, in order to catch up on arrears after years of Labour and Conservative restrictions on public sector capital spending.

A further fiscal advantage of privatization relates to the high inflation context of the 1970s and 1980s, when the combination of rapid price increases and high nominal interest rates created the phenomenon of **frontloading**. With frontloading, the real burden of interest payments on debt comes in the early years of the loan, producing strong, short-term, upward pressure on the charges for debt-financed services. This was a serious problem for the British state housing sector from the mid-1970s (Merrett 1979: 160–66). Water privatization would permit the increased charges necessitated by the substantial investments to raise water quality. But these increases would come from the private sector; central government would be innocent of their imposition and its pockets would be untapped by subsidies to keep public sector water charges down.

The first privatization of the Thatcher years was brought about through the 1980 Housing Act and concerned local authority rented housing (Merrett with Gray 1982: ch. 8). Over the next 15 years, the sale of well in excess of one million of these dwellings was to notch up greater sums of money than all the other privatizations put together. The privatization of water was not seriously considered until 1985 (Evans 1993: 69).

As is noted in the scale economies case study of Chapter 3, in the 1980s there were ten public sector regional water authorities in England and Wales. These authorities, first created in 1974, had responsibility both for water utility and for river basin functions. The most important early decision by the Minister concerned, Mr Nicholas Ridley, was to limit water privatization to freshwater supply and wastewater services, and to hand over to a newly created, public sector, National Rivers Authority the responsibility for pollution control, fisheries, conservation and recreation, some navigations, land drainage, flood defence and management of water resources. Ridley believed that a system which allowed the functions of resource management and pollution control to be combined institutionally with abstraction functions and wastewater disposal was fundamentally flawed. Kinnersley writes of "the dominance of routine and habit in the water sector" and, deliciously, points out that a Tory Minister's wish to restrict the scope of privatization was opposed by the chairmen of the public sector regional water authorities who, in the words of Shakespeare, sought to "out-herod" Herod by insisting that river functions should be privatized too (Kinnersley 1994: 9, 53–4).

Thus, water privatization in England and Wales took the form of the privatization only of the fresh- and wastewater service functions of each of the ten former regional water authorities. For each region a holding company was created which controlled a water utility company delivering the core services, plus any number of "enterprise companies" to pursue non-core service activities. The holding company, as a plc, would offer its shares for sale to the general public. The assets and liabilities of each public sector regional authority were transferred to the subsidiary water service company and it was this core service company that was to be regulated by a newly created government department, Ofwat. Government retained until 1994 a golden controlling share to prevent any of the companies being taken over. The regional boundaries of the big ten and those of the NRA's regions were identical and were derived from those of the 1974 Regional Water Authorities. They constituted entire river basin systems (Manson 1994: 11).

Each of the ten water companies was to hold an appointment, for twenty-five years in the first instance, as provider of water-supply and sewerage services. For each company, there was to be an instrument of appointment, or licence, as it was often called. In essence, this instrument of appointment confirmed the territory in which the company had a monopoly, but it also committed the company to all the regulatory conditions and arrangements that would be applied by Ofwat. As Kinnersley wrote (1994: 63):

> It was essential for this licence to be very robust: the health and life-styles of the whole community could be threatened by a significant failure of water and waste-water services. Therefore, the licence had to provide for Ofwat to have full data on levels of service, and to regulate them as well as water charges. In this, it would have the advice of customer committees which it would organize in each area . . .

The power of the economic regulator had to be secured, in law and through the resources available to carry out its tasks, particularly because the ten regional companies, as well as twenty-one water-only companies, are extreme monopolies. This was of special relevance to water charges, which, even by 1996, in the English and Welsh household sector were 95 per cent derived from a fixed charge on each property. Only some 5 per cent of households paid for their water on a volumetric price basis.

In the late 1980s, major investments were foreseen for the water service companies. Yet there was little prospect of sustained growth in the volume of service output. The clear implication was that charges to consumers would have to rise. Ofwat's powers to prevent monopoly abuse could have been constructed either to control company dividends or to regulate the level of charges. Dividend limitation is relatively simple administratively, but weakens the pressure for efficiency gains. It controls dividends but not costs, and high costs deriving from inefficiency can be passed on to the consumer.

So, Ofwat regulation came to rely on charge-capping, where the company is left to make as much profit as it can, subject to not raising charges beyond specified limits for levels of services, which are also defined and monitored. Regulation is, admittedly, more complex, but the incentive to efficiency is built in from the start, alongside protection against monopoly abuse in charge setting. Under charge-capping, water companies are required to keep the average percentage rise in the fixed charge to customers within a limit which represents the rate of change in the retail price index plus a factor called K. Initial limits were set in 1989 and updated in 1995 (Ofwat 1994a: 5–6).

However, in setting out a legal regime to prevent monopoly abuse, the Conservative government and the Civil Service faced a major difficulty. Once the legislation was in place for regulation of the private water companies, they still had to be floated on the London Stock Exchange. That is, the shares in the companies had to be sold in the market place. If government regulation was thought likely to be too tough, the flotation might be a flop, the market might turn its back on the companies, the shares might remain unsold and the government would have failed to garner in the billions of pounds it had expected to receive from selling off these public sector assets.

In order to avoid such a disastrous possibility, measures were taken. One was to float all ten companies simultaneously, not in sequence, thereby removing the anxiety of water company chairmen about who would be first or last in the queue. Another was the balance sheet reconstructions and debt write-offs that would strengthen the companies' equity positions. A third was to mask existing illegalities in tapwater provision and sewage disposals by the regional water authorities (see below). A fourth was to negotiate with the civil servants within the European Commission to clarify the compliance dates with EC water directives and to insert these in the 1989 Water Act. A fifth was that Ofwat could grant a higher increase in charges than an agreed RPI + K indicated, where a company made a case to the Director-General that investment costs had risen more rapidly than previously anticipated. Finally, the Director-General was required by the 1989 Act to ensure that the companies could achieve an appropriate return on capital to sustain the services they were appointed to provide in their designated areas.

It is argued that the increasing scope and exigence of environmental laws were a source of anxiety to private investors. There can be no doubt there is some truth in this. Certainly, Kinnersley (1994) has exposed what he calls the "environmental betrayal" by ministers and civil servants in obfuscating water pollution measurement, veiling existing illegalities within the regional water authorities' sewage works operations, and lowering quality standards in respect of sewage discharges prior to the publication of the company prospectuses that were required for the flotation.

Yet, from a long-term perspective, the growth in services output for the water companies was likely to come principally from raising quality standards in freshwater supplies and wastewater disposals. As long as the charge-capping regime permitted sufficient profitability for the new rounds of capital accumulation that could deliver these, the water companies would find environmental standards to be the main source of increased turnover. This situation, where private monopolies' turnover growth in real terms was primarily driven by the environmental legislation of a supranational body (the Commission of the European Communities) must be historically unique in the UK.

The Water Act received the Royal Assent in the summer of 1989 and the flotation took place in the following autumn. The plcs were Anglian, Northumbrian, North West, Severn Trent, Southern, South West, Thames, Welsh, Wessex and Yorkshire. The offer price per share was £2.40, with £1.00 payable immediately as the first of three instalments. Institutional investors were obliged to buy packages of shares in all ten companies. The K factor for 1990–95 varied between 3 per cent (for Yorkshire) and 7 per cent (for Northumbrian). The big ten, as well as the twenty-one water-only companies, could set their charge increases below RPI + K in any year and retain a carry-forward potential. On flotation the offer was 2.8

times oversubscribed and on the day dealing began shares traded at 36–57 per cent above the offer price (Kinnersley 1994: 74–9).

Fifteen years ago, the word "privatization" was virtually unknown. But Mrs Thatcher's 1984 flotation of British Telecom proved a watershed. As a *Financial Times* leader put it: "Since then, the systematic sale of state enterprises has accelerated on almost every continent, culminating in the privatization frenzy of eastern Europe and the former Soviet Union." In the UK by 1996, 48 businesses and 950 000 jobs had been transferred from the public to the private sector. Proceeds to the UK Treasury by the end of 1995/6 were some £63 billion.

Plender (1996) argues that in the UK the process has introduced competition, as in telecommunications; disposals have increased public finances; political pressures on industries have reduced, permitting enhanced profitability; and investment plans have been liberated from the constraints on public expenditure. Consumers have been better off, where competition has been introduced. In all industries, save one, real prices have fallen. Efficiency gains have exceeded expectations, although critics argue that a disproportionate amount of the unanticipated gains has gone to shareholders. Plender suggests that the biggest winners are the directors, "many of whom have enjoyed huge pay increases and options amounting to a risk-free punt on shares that were often sold off cheaply by the government." Meanwhile, "efficiency gains have been won largely at the expense of employees, who have been sacked in their thousands".

But, five years after the 1989 Act, Professor John Kay of the London Business School could write: "The privatization of the water industry was, and remains, the least popular and most problematic of utility privatizations" (J. Kay 1994). There were a variety of reasons for this. First, fresh- and wastewater services are an extreme monopoly, for the reasons discussed in Chapter 3, and so there were no benefits from competition. Secondly, as we have seen already, the backlog of capital spending and the EU's environmental legislation necessitated a large capital-spending programme, driving up overhead costs and pushing the rate of increase in charges well above the inflation rate. Thirdly, at a time of high unemployment throughout the EU, thousands of people were sacked from jobs they had looked on as secure, whereas those who remained behind or were newly recruited experienced the disabling stress caused by labour intensification. Fourthly, shareholders rather than consumers were seen to have benefited most from "efficiency gains". Fifthly, the value of directors' salaries, share options and pension rights escalated prodigiously in comparison with the general rate of change in wages and salaries.

In the summer and autumn of 1995, something of a political crisis set in for the water industry. 1995 was a drought year and Yorkshire Water found itself forced to introduce a series of emergency measures to limit its customers' consumption. The Parliamentary Labour Party then made play of the fact that the leakage losses among the big ten ranged between 20 per cent and 38 per cent of the public water supply. For Yorkshire Water it was 36.7 per cent in 1994–5 (Ofwat 1995: 31). Consumer hardships were laid at the door of utility chairmen who were failing to deal effectively with water wastage.

In this climate, regulatory reform appeared on the political agenda, both from Helen Jackson, the Chairperson of the All Party Parliamentary Water Group, and from the quality press (Jackson 1995). In particular, the *Financial Times* suggested that it was desirable to move towards the US model of regulatory commissions, "robustly and visibly independent of government" (*Financial Times* 1995: editorial, 10 March). First, each sector should have a small panel of regulators rather than a single person. Secondly, price formulae reviews should be set at intervals such as five years, but with an additional review permissible when a strong prima facie case existed that, at the most recent review, the utilities had misled the regulator by overstating their costs or understating effective demand. Thirdly, the regulatory panels of all the privatized utility sectors, such as water, gas, telecommunications and electricity, should agree to work from a common set of rules. This would apply, for example, to the definition of depreciation, the definition of capital on which returns are calculated, and the method used to calculate the cost of capital. Fourthly, regulators should be more open about the basis for their decisions and should not shy away from publishing more financial details of the utilities than appear in annual reports. Fifthly, regulators' actions should be scrutinized through a parliamentary select committee on the utilities.

6.11 Final remarks

In this chapter the focus of attention has shifted from the economic analysis of projects to the financial appraisal of enterprises. It is the broader political and macroeconomic context that makes necessary this second perspective. Traditionally, in most countries, the supply of fresh- and wastewater services has been a public sector responsibility. Supply economics was concerned principally with unit costs of output and with the availability of capital funds to finance water projects. Where these funds came from international agencies or the private capital market, in the last resort government was responsible for meeting debt obligations. Simultaneously, the economics of demand was largely ignored, because water, although costly, was not priced. Charges for water use, where they existed, usually took the form of a fixed payment unrelated to volumetric consumption.

However, with the global shift in the past 15 years to the privatization of public sector assets, the financial viability of water enterprises has become a matter of great interest. In some cases, this was simply because such enterprises were privatized. In other cases, where water remained in the public sector, a clear intention was expressed by national governments and international financial institutions that, at least, such public sector companies should become "autonomous", that is, financially self-sufficient. In this second case, once again, the financial analysis of the enterprise took on great importance. In this volume, the case studies from Latvia, Peru, San Serif and the UK all illustrate these changing relationships. Within the framework of financial accounting for enterprises, the keenest regard has been to the ability of water companies to repay new loans taken up for the rehabilitation or expansion of hydrosocial infrastructures.

This chapter provides a brief introduction to enterprise accounting using law and practice in their application to UK public limited companies. Financial accounting is defined, the variety of separate accounts is reviewed, double-entry bookkeeping is presented, the fundamental accounting identity is set down, and the trial balance and the adjustments it requires prior to the presentation of the basic financial statements are described.

A description is given next of the profit and loss account. Profit (or loss) is defined as the difference between current income and current expenditure, with respect to the economic activities of a given financial year. The importance of subtracting an item for the depreciation of fixed assets is stressed, and similarly for bad and doubtful debts. For water enterprises, gross profit is the income from the trading account less the outgoings on the processing account. The latter may provide the basis for the separate pricing of fresh water and of wastewater services. The main difference between gross and net profit is likely to be the interest charges on the company's borrowing. Net profit is then split between the taxman, the shareholder and retained profits.

The importance of debt interest then leads on to consideration of a company's six distinct forms of gaining command over resources. These are share capital, debentures, loans, overdrafts, trade credit and retained profits.

The next financial statement considered is the balance sheet, where fixed assets plus current assets less current liabilities is equal to **equity** plus long-term loans. The third and last of the main financial statements is that of the cashflow, which describes in detail the movement in an entity's cash position during a defined period.

Next the chapter sets out the problem of leverage for bias, which financial accounting and project analysis encounter. In footballing terms, the referee is offered inducements (or receives threats) to persuade him to favour one result rather than another. Accountancy's ethical conventions of consistency, objectivity, relevance and prudence are described, with an illustration of how they were breached in a case where the World Bank was the "shadow client".

In §6.8, a gap left by the coverage of Chapter 5 is filled, with a review of five distinct techniques used in the private appraisal of projects: rate of profit on capital, payback period, net present value, internal rate of return and the discounted profit:capital ratio. In my view, the last of these is the best. A brief exposé of the risks associated with hydrosocial investment suggests that, in spite of its natural monopoly position, these infrastructures embody substantial financial risks for private enterprise because of their inflexibility, their localized market (the immediate area constitutes a **monopsony**), and the ever-present possibility of shifts in government regulation of output quality, abstraction charges, discharge fees, and the price of water.

The first case study is taken from Boca in San Serif, where the water utility in 1995 was close to the social service type of Chapter 4. The politics of awarding this feasibility study contract are indicated. The key tabulations of profit and loss, balance sheet and cashflow are presented. Some of the weaknesses and strengths of the financial work are highlighted, including an implicitly *positive* value for the price elasticity of demand. Lastly, internal and external politics intrude again in explaining why enterprise accounting had completely displaced project analysis.

The second case study is of the privatization of the water utilities in England and Wales. The political context of this shift from a social service to an intermediate institutional form are set out, both ideological and fiscal. Attention is drawn to the separation of environmental from economic regulation. The legal autonomy of the core service from enterprise activities is set out, as well as the licensing arrangements, and the status and role of the water regulator. The powerful interrelation between the legislation to regulate the core company and the market flotation of the holding company is demonstrated. Politics is shown to intrude on financial and environmental decision-making prior to the share sale. Paradoxically, the EU's environmental directives are judged to constitute both a source of risk *and* the driving source for turnover growth. Lastly, the recent travails of the privatization in England and Wales are dissected and the necessary revamping of the regulatory process is addressed.

This chapter began by citing the view of the EPDRB task force that, by maximizing private sector involvement in fresh- and wastewater services, central government is relieved of the "burden". This is an appropriate point to review such a standpoint. To what degree, then, is public sector provision of hydrosocial services a *financial* burden on central government? To answer that question, our earlier distinction between capital and current expenditure will be useful, and these two approaches will be taken in turn. So, again, is public infrastructure expenditure on *capital* account a financial burden? At first view, the answer is – of course not. Raising funds on the national and international capital markets is central to the forward movement of market economies in securing the provision of collective goods and of technological innovation in industry. But, to continue, is it the case that, precisely, it is government getting and spending that threatens the macroeconomy, because of its inflationary impact or because it creates a public sector borrowing requirement?

In fact, there is nothing inherently inflationary about capital expenditure, either by government or by the private sector. A steady stream of public (or private) investment projects certainly contributes to the effective demand for labour of all kinds, materials and capital goods but, provided total saving in a country is high, there is no inherent reason why such steady state demand should be inflationary, even in a fully employed economy. Only in the case where there are supply-side bottlenecks, especially in the labour market, because of excess demand caused by inadequate national savings patterns or by an undervalued exchange rate, would infrastructural investment reinforce inflationary trends. But that is as much true of the private as of the public sector.

Leaving the real economy aside, is it then the case that *public* investment, by adding to the public sector borrowing requirement, drives up the money supply with inflationary impacts? Here, we should point out that the provision of financial and wastewater services in the public sector can be, and usually is, the responsibility of national, regional or municipal public trading enterprises. As Will Hutton points out, the internationally agreed system of national accounts *excludes* borrowing by such enterprises from the definition of general government borrowing and defines the key financial target for government in terms of the general government financial deficit (Hutton 1995). The general government financial deficit excludes public trading enterprise spending from its calculations.

Let us ask, then, is public infrastructure expenditure on *current* account a financial burden? There are two possibilities here. First, one can take the case where a well run public trading enterprise in the water sector covers its current spending obligations (including loan interest, of course) from the charges it levies on users. Here there is no inflationary impact.

Secondly, let us suppose that the public trading enterprise's management has been appointed within a general climate of corruption in the public sector, or receives low salaries because of government policy on the public sector wages and salary bill. Here, we can expect gross managerial incompetence, a poor service to consumers, low capacity utilization, high leakages, serious overstaffing and high unit costs – the complete catastrophe. In this second case, charges to the consumer will be higher or the public trading enterprise will have to rely on central government subsidy, or both. Such a situation is most definitely a financial burden for central government.

In summary, public trading enterprise provision of water services is not a financial burden for central government except in cases where public sector management is grossly incompetent, and where no reasonable and immediate prospect exists for reversing this situation. In such cases, the argument for privatization is strong indeed. Where public trading enterprises have a strong managerial track record, or good arguments exist that a reformed or new public trading enterprise will be well run and will invest unconstrained by public sector borrowing-requirement targets, the argument for privatization is weak. Evolutionary political economy eschews policy recommendations of a completely general nature. The discussion of privatization or socialization should be pragmatic, and proposals should be contingent upon the specific circumstances and history of a region or country. Key areas to address are the impact of institutional change on the quality of service to customers, the long-term environmental outcomes, the wise use of public money, and the wellbeing of the utilities' workforce.

CHAPTER SEVEN
Water resource planning for a sustainable society

7.1 Introduction

Chapter 5's enquiry into evaluation methodology explicitly postponed consideration of projects' environmental benefits and costs. In correcting this omission, it would be a mistake immediately to return at the level of the *project*. Projects, by their nature, are spatially delimited. In addition, project analysis is usually confined to relatively short periods of time (5, 10, 25 years) and the weight of later events, as we have seen, is powerfully reduced by discounting procedures. A prior task, in examining the interdependence of the economy and the environment in respect of water, is to develop a synoptic and long-term perspective. To use the language of Johan Åkerman, whereas the "time horizon" of project planners is short- and medium-term, the time horizon for society in respect of environmental change must be secular and the scale must be global (Åkerman 1960; Mjøset 1994: 10–11).

To carry through such a task, the concept of sustainability is of the most vital significance. However, it has to be admitted that, in the mid-1990s, among the throngs attending conferences, seminars and workshops on sustainability and water – or energy or nature conservation or transport or land-use planning – one could not doubt the existence of real confusion and disagreement about the meaning of this concept (Merrett 1995). For the naïve, this uncertainty might have been dispelled by a conference chairperson confidently stating that the Brundtland definition would be used. "Sustainable development is development that meets the needs of the present without compromising the ability of future generations to meet their own needs" (WCED 1987: 43). But others asked themselves why it is that the adjective "sustainable" is attached by one author to "environment" and by another to "development". Perhaps, since sustainability is seen to be so praiseworthy, each writer or speaker simply hooks sustainable on to the noun that expresses the activity or entity that person most values. These philosophical qualms are reinforced among those who know that sustainability in its current sense is first to be found in *Blueprint for survival*, where an early reference is made to "a sustainable society" (Goldsmith et al. 1972: 2).

Unfortunately, there can be no doubt that "sustainable development" is now the dominant term. Yet many regard development as essentially an economic category, hardly distinguishable from economic growth. After all, it was Gro Harlem Brundtland who wrote in *Our common future*: "What is needed now is a new era of economic growth – growth that is forceful and at the same time socially and environmentally sustainable." (WCED 1987: xii). Thus, in a real sense, economics has attempted to appropriate the concept of sustainability for itself, when it is clear that the core relationships in question are far broader than that of any single discipline. Neoclassical economics has always been the most imperialist of the social sciences.

In spite of all these ambiguities and confusions, sustainability is here to stay. Moreover, the concept's relevance to water resource planning is powerful and will be permanent. What is needed, in these early days of a quite new dimension to an old profession, is to put sustainability into its historical context, and to derive from that understanding what should be *the principal fields of action* in the sustainable management of water resources.

7.2 Power to alter the nature of the world

An appropriate starting point for an overarching perspective on sustainability, environment and development is the economic changes that first began to take place in Britain in the eighteenth century and which are known as the Industrial Revolution. Over a period of decades, British entrepreneurs found new ways of organizing the work process in the manufacture of commodities, based on the division of wage-labour within factories. With the profits made in domestic and overseas markets, industrialists financed the purchase of new machinery embodying technical innovations that enormously increased labour productivity in specific phases of the manufacturing process.

So, finance, work process organization, wage-labour, market expansion and technological innovation all came together to constitute a new mode of production, of which Adam Smith was the first and greatest prophet (Smith 1961). Capitalism spread across the globe in the next 250 years, often with the assistance of governments to their own national industries, and embracing service outputs and agricultural production as well as manufactures. Concomitantly, there developed new marketing techniques by which business sought to secure the sale of its products on a mass scale. Fundamental human needs, such as those for friendship and sexual love, were harnessed by the advertising industry to sell commodities.

The capitalist mode of production had an almost miraculous power to increase the measured gross domestic product of nation-states. At the same time, the exponential growth of food output, the development of public health infrastructures and the advances of medical science all led to an historically unprecedented growth in world population. As Roberts (1990: 908) has said, commenting on a world population exceeding five billion persons: "Though it had taken at least 50 000 years for *Homo sapiens* to increase to 1000 millions (a figure reached in 1840 or so) the last 1000 million of his species took only 15 years to be added to a total growing more and more rapidly." This expansion in population numbers and in the rate of economic activity per head – in the sense of production *and* consumption per capita – had powerful effects on the natural world.

For the economy, the biosphere exists as a resource, sink and servicer. As provider of *resources* for the processes of production and consumption, it offers three gorges for a dam, a forest cleared for a city, a delta for eco-tourism, a desert landscape for an interstate highway, a lake for fishing, a swamp drained for agriculture, water for the mill, coal to the furnace, ore to the smelter, wood to the logger, game to the table and, for the apothecary, the bones of the tiger.

As a *sink* for the dross of economic activity, the biosphere offers the atmosphere for chlorofluorocarbons and oxides of nitrogen, the land for the city's tide of refuse and the Earth's waters for every effluent and waste from agriculture, industry and the domestic sector.

Above all else, the biosphere and the solar system provide the *life-support services* without which neither production nor consumption nor any form of life on Earth would be possible: the atmosphere as a source of oxygen, protection against ultraviolet radiation, and shield from meteoric bombardment; the water stocks and flows of the hydrological cycle, with its role in photosynthesis, natural desalination and climate modulation; the Sun's light and heat, the source of eternal energy; and gravity itself, without which all things fall apart.

Most resources and sink locations, with the partial exception of the oceans, can be physically or legally appropriated to specific actors for their value in the economic process. In contrast, the biosphere's life-support services are beyond private or state ownership. What scarcities would emerge and what fortunes would be won if Enterprise Incorporated, with a dictatorship in its pocket, could monopolize the air, light, heat and gravity!

To recapitulate, since the eighteenth century, there has been a long and powerful expansion in both the global population of the species *Homo sapiens* and in its per capita rate of making and using goods and services. This demographic and economic dynamic, periodically interrupted by destructive wars of a new kind, has brought with it ever greater stress on the natural world as a provider of resources, as a sink for wastes and as a life-support envelope.

As a result, one can distinguish four types of negative impact of the world economy on the global environment: the destruction of species and habitats, the depletion of non-renewable resources, the degradation of natural and built environments, and the disabling of life-support services, where the ozone holes and global climatic change are particularly life-threatening. I shall refer collectively to these four processes – destruction, depletion, degradation and disablement – as **inquination**.[1]

---

1. No single word exists in the English language that groups together as a single concept these four processes of destruction, depletion, degradation and disablement. One is needed, and "inquination" has an adequate pedigree: defined in the *Oxford English dictionary* (Allen 1990: 1444) as "The activity of polluting, defiling, or corrupting . . .", it derives from the Latin *inquinare* (to pollute, etc.).

As Rachel Carson said (1987: 5):

The history of life on Earth has been a history of interaction between living things and their surroundings. To a large extent, the physical form and the habits of the Earth's vegetation and its animal life have been molded by the environment. Considering the whole span of earthly time, the opposite effect, in which life actually modifies its surroundings, has been relatively slight. Only within the moment of time represented by the present century has one species – man – acquired significant power to alter the nature of his world.

In the command economies, such as the Soviet Union after 1917, the combination of industrialization, and the suppression of citizens' movements which was a characteristic of "democratic centralism", had even more damaging environmental consequences than those in the capitalist heartlands.

Inquination has provided the material basis for the gathering strength of environmentalism, particularly since the first publication of Carson's *Silent spring* in 1962. Tellegen and Gustafsson, for example, demonstrate this in their outline histories of these social movements in the twentieth century, for the case of the Netherlands and Sweden respectively (Tellegen 1981, Gustafsson 1993).

What has been the driving motivation of these new political forces? In respect of the destruction of the natural world, in part it has been the *instrumental* value of ecosystems and their populations to human society. Yet a recognition of the *intrinsic* value of nature has also contributed to environmental campaigns (Simmons 1993: 118–37). Opposition to the destructive powers of economic activity has been buttressed by a much wider understanding of the beauty and complex symbioses of the natural world, with the impact of magnificent television programmes such as those of the BBC Natural History Unit from Bristol.

Environmentalism's phenomenal energy also derives from the degradation of the air, water and land of rural and urban spaces by industrial pollutants and wastes (Kapp 1983). Human societies value highly the quality of residential, work and leisure environments and, if they feel they have the power, will resist their being poisoned.

So – motivated by their social valuation of the natural world and the quality of the environment – educational, campaigning and overtly political movements have shifted the legislation and policies of government in respect of the environment. But the changes we have witnessed in Europe and other continents since the mid-1980s have a third source.

Inquination now constitutes an emerging **contradiction** for the market system on a world scale, just as the growth of the great industrial cities created contradictions within mid-nineteenth century Europe (Merrett 1979: 1–8). The term contradiction refers to the development of forces within a social formation, which threaten the long-term expansion and stability of that society – in effect, which endanger its sustainability. The chronic, corrosive processes of modern capitalism threaten to undermine the foundations of its phenomenal material triumphs.

The destruction of habitats is associated with desertification and flash-flooding, for example, with their social and economic costs, as well as the loss of the rain-forests' capacity to absorb atmospheric carbon dioxide. The depletion of fossil fuels and mineral ores is likely to render industrial processing less efficient and therefore make it more costly. Environmental degradation resulting from the pollution of the air, water and soil creates not only the resistance movements of environmentalism but lowers the productivity of the labour force. The disablement of the global ecosystem threatens production and consumption processes at the macro-scale, through global warming and large-scale climatic change.

These patterns of change suggest that the GDP of nation-states in the twenty-first century will be given over more and more to coping with the negative environmental externalities of the market system. These forms of damage limitation include the treatment of cancers, flood protection, population resettlement and decontamination measures after nuclear or chemical accidents. For the next century, humankind's principal concern will not be the limits to growth, but growth's destructiveness of the environment.

7.3 The meaning of sustainability

It has been argued above that the general development of environmental policy since the mid-1960s, and the quest for sustainability since the late 1980s, derive from the values human actors place on other species and on habitats, from the values we place on the quality of natural and built environments, and from a growing sense of a serious, long-term malfunction in the dominant mode of production and consumption.

This suggests that Goldsmith and his collaborators in *Blueprint for survival* were right to apply the term "sustainable" above all to *society*, in the sense of both the nation-state and to human society at the global scale. If this is accepted, an appropriate, generic definition of sustainability is as follows: a sustainable society is one in which, for the indefinite future, human communities sustain and regenerate the species and habitats of the natural world, sustain and rehabilitate the quality of natural and built environments, sustain the global ecosystem's power to provide life-support services, and sustain and transform society's economic capacity to meet the material and cultural needs of its people.

In a nutshell, the sustainable society refers at the global scale to the sustainability of nature, of built environments, of the life-support envelope and of production and consumption. Sustainability is a key concept, both of political economy and environmentalism. No single discipline or profession should arrogate the idea to itself. Within this broad framework, it then becomes meaningful to locate much more specific terms, such as sustainable development, sustainable region and sustainable transport policy. But without that wider structure of meaning, terms such as the sustainable city have no sense. For sustainability is a holistic and protean concept with a worldwide scale of reference.

7.4 The role of water resource planning

Governmental response to the environmentalist challenge and to the long-term threat to the market system has taken many forms. In some cases this has been to regulate the production of pollutants and wastes. Some countries have seen the use of pricing policies, taxes and subsidies as an effective way of modifying economic behaviour and therefore environmental impacts. Recent proposals in Europe for an energy tax and a carbon tax spring to mind. Elsewhere, technological innovation in urban infrastructures are promoted, such as the use of combined heat and power in Denmark and Sweden (Månsson 1992). Calls are made for shifts in agricultural policy, to encourage biomass production for energy purposes (Webb & Gossop 1993). Within a quite different perspective, there have been powerful voices, such as that of Arne Naess (1989), which seek nothing less than the transformation of society itself. Furthermore, at the United Nations Conference on Environment and Development in Rio de Janeiro in 1992 and, even more forcefully at the 1994 Cairo conference on world population, participants saw as central to a global strategy the promotion of family planning to limit population growth (UNCED 1992).

At the sectoral level, alongside transport and energy policy for example, there can be no doubt that water resource management is one of the principal keys to unlock the gate to the sustainable society. As Söderbaum (1994) insists, this must be a concern of the whole of society, including government, businesses, NGOs and individuals. Sectoral policies and their implementation are far too important to be left to government alone.

Yet, it still remains here to propose how water strategies can contribute to this goal. The *Oxford English dictionary* gives one of the meanings of the word *principle* as "a fundamental truth or law as the basis of meaning or action" (Allen 1990). I propose the following sustainability principle for water resource management:

Water resource planning for a sustainable society should seek:
- the protection of water's hydrocyclical capacity to renew its ground- and surface-water flows and stocks
- the purification of water from domestic, agricultural and industrial effluents
- the conservation of society's species and natural habitats in all their fresh- and coastal-water environments
- the husbandry of water in its supply and use
- the supply of fresh water sufficient to meet the biological, cultural and economic needs of society's human populations
- the drainage of stormwater, and the protection of rural and urban communities against floods.

As can be seen, the sustainability principle embraces six fields of action. Examples of the necessity for such action are set out below.

Protection of the hydrocycle

Discussion of environment–economy interdependence in recent years includes references to the second law of thermodynamics, also known as the entropy law. In Khalil's words, this law (1994: 186):

> simply stipulates that in an isolated chamber of gas, temperature or pressure would tend to spread uniformly throughout. The arrival at such an equilibrium state, given the total energy and size of the chamber, maximizes the magnitude of a mathematical parameter, viz., entropy.

Authors such as Georgescu-Roegen and Boulding have argued that the economy extracts low-entropy energy and returns to the environment high-entropy wastes (Foster 1994: 26).

The contrast between water and fossil fuels as energy sources is strikingly apparent in this respect. In the course of the hydrocycle, the processes of evapotranspiration, condensation, precipitation and gravity-flow constantly gather together again the H_2O molecules dispersed from the land through runoff to the oceans. This cycle is powered by solar and gravitational energy. As a result, water is the paradigm case of a renewable resource. For this reason, too, hydroelectric power and tidal and wave power generation are all regarded as renewable energy resources.

However, at specific spatial scales, the hydrocycle can be disabled. This occurs when extreme quantitative stress is imposed at the local level (see Eq. 2.9). In such cases, aquifers are exhausted, streams and rivers dry up, and lake levels fall dramatically.

McDonald & Kay (1988: 50–53) provide a most striking example of such a process in the case of groundwater mining of the Hueco Bolsón aquifer beneath the Río Grande, where it forms the border between Mexico and the USA. Abstraction to meet irrigation use led to a fall in the water level by 25 m between 1903 and 1976 and, in the mid-1970s, the depletion rate was running at 1 per cent per year. The extreme stress on the resource led to increased pumping costs, groundwater contamination from saline and polluted sources and subsidence.

Purification of water

Water treatment, before or after use, is of course a mainstream utility function, and has been discussed at length in Chapters 3–4. The question raised by Kinnersley (1994: 2) concerns the quantitative relationship between the hydrological and the hydrosocial cycles:

> Now we understand the medical and scientific implications of good water quality, and have the technology to deal with most of the hazards of which we are aware. But can we cope reliably with the hugely increased volumes of effluents and wastes that modern communities want to inject into inland and coastal waters? Can we even measure how much pollution is now reaching rivers or underground aquifers from diffuse sources such as accidents to

road tankers carrying chemicals, the use of pesticides and fertilizers, and other industrial or agricultural procedures regarded as necessary nowadays to achieve greater efficiency or productivity almost regardless of their consequences?

A contemporary example of this challenge comes from the People's Republic of China. Anhui province is a land-locked area several hundred kilometres to the west of Shanghai and is a major centre for a wide range of industries. Much of the province's economic activity is located in the Lake Chao basin, which contains 3000 factories and 6 million people. The area of the Lake's watershed is 9258km^2, and the lake itself has a surface area of 760km^2, but it is only 2–3 m deep.

Lake Chao is the fifth largest lake in China and is the main source of abstracted water for cities and industries in the watershed. It also supports fishing, agriculture and tourism. In 1995, 40 million m^3 of untreated domestic sewage and 160 million m^3 of untreated industrial wastewater was discharged into streams and drains that flow into Lake Chao a few kilometres down stream of the urban centres. Moreover, deforestation and defoliation have caused increased soil erosion, so that the lake is silting up. Runoff from agricultural land is rich in nutrients and other contaminants.

Unsurprisingly, the quality of water in Lake Chao has deteriorated. Algal blooms increase the cost of water treatment and, because they can plug the water treatment plant intake, have sometimes deprived Hefei City, the major urban centre, of water for several weeks at a time. The health of urban residents is threatened and the **eutrophic** condition of the lake's water reduces fishing yields. The continuing economic development of the basin is expected to increase present levels of wastewater from the domestic and industrial sectors by 60 per cent and 40 per cent respectively by the year 2000. It was this situation that led the Asian Development Bank to launch in 1993 several technical assistance projects to strengthen improved treatment of the basin's wastewater.

Conservation of species and habitats in fresh- and coastal water environments
The Global Environmental Facility is a cooperative venture among national governments, the World Bank, the United Nations Development Programme and the United Nations Environment Programme. It supports actions to reduce and limit the emission of greenhouse gases, to preserve the Earth's biological diversity and natural habitats, to control the pollution of international waters, and to protect the ozone layer. The GEF provides grants and low-interest loans to eligible countries and also supports international environmental management and the transfer of environmentally benign technologies. Usually a project must not be economically viable without support from the GEF, and most funding will be provided for investment projects. Other supported activities include technical assistance, pre-investment and feasibility studies, scientific research, and training.

The task force of the environmental programme for the Danube river basin has described a GEF project for the Danube delta in Romania and the Ukraine (EPDRB task force 1995: 56, 65):

(The delta) consists of rivers, lakes, reed swamps, meadows, sand dunes and forests. It is a rich economic resource of fish, timber and reed and is home to about 80 000 people. It is important as a national tourist centre and has a considerable potential for ecotourism . . . The major impacts on the delta eco-systems result from the changes both in upstream conditions, such as the manipulation of water discharge and pollution, and from changes in the delta itself. The most significant activities in recent decades have been the creation of a canal network in the delta to improve access and the circulation of water through the delta, and the reduction of the wetland area by the construction of agricultural polders and fishponds. As a result, biodiversity has been reduced . . .

GEF funding aims to promote the rehabilitation of wetland areas for the purpose of restructuring the delta's biodiversity and increasing its nutrient filtering capacity. The pilot project experience could be used in other attempts to reconstruct the ecological functions of wetlands both in the delta and up stream in the Danube basin.

Husbandry of water in its supply and use
Water husbandry embraces the reduction of supply losses discussed in Chapter 2 and the demand management material of Chapter 4.

The example here is taken from the Middle East, which faces an environmental crisis, much of it as a result of water scarcity. Within the region, Israel is regarded as having made most progress in its husbandry of water, although the much greater consumption of water per capita among the Jewish population, in contrast to the Palestinians, is striking.

As Arlosoroff (1995) writes, water use in Israel operates within a tight allocation system and relatively high prices, although irrigation water supplied by the National Water Company receives subsidy. In 1995, farmers paid 50 per cent of their frozen 1989 allocation at US$0.15 per m^3, US$0.19 for the next 30 per cent, and US$0.25 for the rest. A formula exists for the automatic updating of these rates. A comparison of irrigation water prices between Jordan and Israel is said partially to explain the gap in agricultural output per m^3 of water consumed.

Urban water is not subsidized. City utilities pay costs at gate and then add distribution costs, sanctions for higher unaccounted-for-water, and effluent charges. Including sewerage charges, the marginal rate for many households is US$1.2 per m^3.

In addition, Israel has done much to reduce supply losses, vigorously practises external re-use, grows cotton and other row-crops with automated drip irrigation systems, promotes domestic water conservation kits, and new industries are installing water-efficient cooling apparatus (ibid. 1995).

Freshwater supply for the domestic, agricultural and industrial sectors
In setting out the characteristics of a sustainable society (see §7.3), the last of these was to "sustain and transform society's economic capacity to meet the material and cultural needs of its people". Thus, the fifth field of action in water resource planning for sustainability is the supply of fresh water to households, agriculture and industry. The supply side was the focus of Chapters 2 and 3, where some 13 options available for increasing freshwater supplies were set out. This fifth field of action, as an engineering process, should be guided by criteria drawn from the first four fields: freshwater supply should operate within the limits of the hydrocycle, should not itself cause environmental pollution, should be consistent with prevailing conservation policy, and should be pursued in tandem with demand management strategies.

Flood protection and land drainage
Flood protection and land drainage are services closely akin to freshwater supply, in that they concern society's capacity to meet the material needs of its people. The example here is taken from the Thames estuary, which is vulnerable to flooding. The height of high tide in central London is rising at 75 cm per century, because of the slow, downward tilt of South East England, to London's sinking in its bed of clay, and to rising sea levels. Even more potently, an extended trough of low atmospheric pressure moving into the British Isles from the Atlantic can raise the level of sea water below it by up to 9 m. If this water plateau passes into the North Sea, it can cause excessively high surge tides. Such tides, especially if they coincide with the twice-monthly "spring" tide and heavy rain in the Thames catchment, could inundate London. The central feature of the capital's defence is the Thames Barrier, stretching 520 m across the river at Woolwich. Closure of the Barrier takes only 30 minutes and can hold back the incoming tide completely (NRA Thames Region undated).

7.5 Case study: nature conservation in the UK

In the UK, societies concerned with the conservation of nature and the built heritage are astonishingly varied and numerous. They collectively enjoy large membership and are politically important, at both the national and local levels because of the property they own, the ability to harness the support of their members and the professional expertise they wield.

One of the most important of these NGOs is the Royal Society for the Protection of Birds (RSPB). The RSPB's principal objective is to protect the wild bird species of the UK and the habitats within which they flourish.[2]

In 1995, the Society published *Water wise: the RSPB's proposals for using water wisely*, focusing "on the quantity of water needed to maintain wetlands and on how our precious water resources can be managed sustainably" (Fowler 1995: 3), and this document is the source of this chapter's first case study.

The report uses ten categories to classify UK bird habitats. These are estuarine (both intertidal and saltmarsh); swamp, fen and carr; lowland wet grassland; upland bog; native pine; upland broadleaved mixed woodland; oligotrophic and mesotrophic waters; eutrophic waters; rivers and streams; and pits and reservoirs. For each category, it shows the potential impact of the water resources sector on sites and on birds.

Negative site impacts are reduced freshwater input, increased saline intrusion, change in sedimentation, lowering of the water table, drying out of site, reduced winter flooding, flooding by impoundment, increased fluctuation in water levels, low river flows, and increased regulation of habitat. Negative bird impacts are alteration in range and density of plant and invertebrate food sources, shortening of the breeding season, and deterioration, reduction and removal of habitat.

For example, Loch Tarff in the Highland Region of Scotland is a Special Protection Area designated under the EU's 1979 Birds Directive (CEC 1985). But it is also the reservoir for Fort Augustus, where natural fluctuations in the water level are exacerbated by changes caused in the reservoir's operational management. The artificial changes in level happen frequently and are regarded as a serious factor in the reduced breeding success of Slavonian grebes and black-throated divers at the site. Water-level fluctuations can leave nests stranded above the water's edge, making eggs and chicks vulnerable to predators (Fowler 1995: 6).

In respect of water resource planning for a sustainable society, the orientation of the RSPB work is, of course, to the third field of action of §7.4, the conservation of species and habitats. In contrast, in developing its policy recommendations in respect of water, the main focus is on fields of action one and four: the protection of water's hydrocyclical capacity to renew its ground- and surface water flows and stocks; and the husbandry of water in its supply and use. The fact that *Water wise* concerns itself with only the *quantity* of water prevents it considering policy in the third field of action, the purification of water from domestic, agricultural and industrial effluents. This is a serious weakness in the scope of the document.

Policy for protection of the hydrocycle is broad in scope and covers site protection, improved regulation and financial measures. The focus is abstraction practice and policy. It was the 1963 Water Resources Act that first introduced abstraction licensing to England and Wales, and in 1995 all abstractions equal to or greater than $20\,m^3$ per day were licensed by the National Rivers Authority. However, many of the licences are inherited from those given by right to existing abstractors in 1963, regardless of whether the abstraction detrimentally affected river flows or

---

2. The Society is the largest wildlife conservation charity in Europe, with over 890000 members in 1995 and around 800 staff. The Young Ornithologists Club, its junior membership section, has over 130000 members and is the largest wildlife club for young people in the world. *Birds*, the members' magazine, has a readership of 1.8 million. The RSPB's 130 reserves cover 83000 hectares and receive over a million visitors each year. Annual income in 1994–5 was £34 million, the bulk of which came from membership subscriptions, legacies and fundraising. In the same year £20 million was spent on direct conservation expenditure. In 1993, the Society joined with bird and habitat organizations worldwide to form a global partnership called Bird Life International.

sustainable aquifer yields. The NRA may revoke or revise licences, but funds for compensation have to be found from its regional water resources budget.

In Scotland, private water supply requires no consent to abstract. Abstraction for the public water supply and by electricity-generating companies requires a government Water Order. For irrigation, River Purification Boards must first seek from government a control order, which then permits a licensing arrangement. In Northern Ireland, no comprehensive abstraction licensing exists, although abstractions for industrial use require consent from the Water Executive. As in Scotland, legislation does not allow for charges to be made for abstractions.

For future policy in the UK, the RSPB recommends that water resources management should incorporate formal consultation with statutory and non-statutory bodies on proposed water resource development, full assessments of the potential environmental effects of any proposed water development, and security for wildlife and habitats by including a duty to further nature conservation.

In England and Wales, the RSPB points out that a philosophy of balance between the needs of the environment and those of the abstractors has led to "piecemeal destruction and attrition of extremely important nature conservation areas". It is urged that, for site-specific cases, priority should be given to existing environmental obligations.

The Royal Society for the Protection of Birds proposes that the Environment Agency, created in 1996, should implement a rolling programme to review the sustainability of water use in all catchments. The key quantitative questions are: How much water is present? How much water is needed to protect wildlife and their habitats? How much is left for other purposes? The review should give priority to abstractions known or likely to affect adversely a proposed or designated site of nature conservation importance. The Agency should commit itself to revoking or reducing abstractions where necessary, and central government should make funds available for this, if the NRA budget is insufficient. Moreover, the Environment Agency should introduce time-limited licences, to avoid future problems of licence revocation.

In Scotland and Northern Ireland, the RSPB believes abstraction control should be introduced for all abstractions and all areas. A comprehensive register of all abstractions and impoundments is required, as well as a programme of regular surface- and groundwater monitoring. A rolling programme of abstraction review must be implemented. Government authorities should be given a duty to further and promote nature conservation, and abstraction control should not introduce any licences of right.

With respect to abstraction charges, as was discussed in §3.2, all major abstractors in England and Wales make payments based on the total quantity of water they are licensed to take. "These charges are set to recover the NRA's expenditure on water resource activities. As a result, charges are generally too low to have any significant effect on water using behaviour." (Fowler 1995: 17).

Water wise recommends the use of incentive charging for abstraction, as is already practised in Germany and the Netherlands. These arrangements vary the charge according to the volume of water taken, the location of the abstraction, the time of the abstraction, and the amount of water likely to be returned directly to the water course. Interestingly, the RSPB seems to doubt the usefulness of environmental valuation practices: "Charges should be set at a level needed to protect the environment by influencing abstractors' behaviour rather than reflecting the exact environmental cost of the abstraction, which in most cases will be difficult to quantify". This issue is discussed in Chapter 8.

In Scotland and Northern Ireland, where no abstraction charges are payable, abstraction charging will provide the funding for the sustainable management of water resources and should be incorporated into the new control structures for those two areas.

The RSPB also considers the possible role of tradable abstraction rights (see §3.7). In this case, a fixed volume of surface and groundwater is first allocated among competing abstractors in a river basin. Then a formal market is created where these abstraction permits can be traded between users. So, the regulator sets environmental limits to total abstraction but then creates a market for its final allocation. Where the system begins with aggregate over-abstraction, the government can buy back its permits over time.

At least three difficulties are identified for the UK. The first is that timing and location of abstraction, after trading, may breach environmental requirements. In that sense, abstracted water cannot be considered a homogeneous good from a societal point of view. Secondly, trading may reallocate take-up from abstractors who recycle their supply back into water courses to those who do not. Thirdly, permits may be reallocated from "sleeping" licensees to users who will make full use of them.

The last hydrocyclical issue to be covered here is the RSPB's views on flood defence management. It is argued that a floodplain acts like a sponge in its capacity to store floodwater. However, the UK's flood defence practices have altered the natural regime of the floodplain, such that flood water drains more quickly off the land, cutting its storage functions. The Society argues that, where practically possible, these policies should be reversed. This would reduce downstream flooding, improve conservation interest, raise the water table in some aquifers, help check the problem of low river flows, and bring water quality improvement.

The RSPB's three key recommendations in the protection of the hydrocycle – the first field of action for sustainable water resource planning – are:

- *Site protection* The statutory nature conservation agencies should undertake a comprehensive review of the adverse impacts of water abstractions and impoundments on recognized important wildlife sites. Where such a review has been completed, actions must be prioritized and taken to remove these impacts.
- *Improved regulation* A consistent regulatory system should be operated to control water abstraction throughout the UK. Protection of the environment should be a key consideration in such a mechanism. This system should be based on environmentally acceptable flows, that is, flows able to sustain biodiversity. Proper assessment of groundwater availability is essential.

• *Financial measures*   Greater use should be made of financial measures in the management of water resources to encourage the efficient long-term use of water, and to ensure that the price we pay for water reflects the environmental cost of removing that water from the environment.

7.6 Case study: Californian agriculture 1990–2020

In 1995 California had a population of some 33 million people. Gleick et al. (1995), working from the Pacific Institute for Studies in Development, Environment, and Security, published *California water 2020: a sustainable vision*. This provides a remarkable overview of the currently unsustainable supply and use of water in the state, sets out a new philosophy for the future, describes how water consumption can be reduced, and details implementation practices to attain the authors' vision.

Gleick and his colleagues are deeply concerned about the present state of affairs in California. In many areas, groundwater is being used at a rate exceeding that of natural replenishment, causing land to subside and threatening some aquifers with collapse. Groundwater use is almost entirely unmonitored and uncontrolled, hindering rational management. Urban water use is inefficient. Agricultural policy encourages the production of water-intensive low-valued crops, and farming run-off is contaminated by chemicals. The water needs of the natural environment are not well understood and are rarely met. Fish and wildlife species are being driven towards extinction, and habitats are being destroyed by water withdrawal and urban development. "Yet those responsible for managing and protecting the state's freshwater resources continue to plan on the basis of outdated and inappropriate assumptions." (Gleick et al. 1995: 99)

This case study is drawn from *California water 2020*, specifically in respect of the agricultural sector. In 1990, the farming industry, with a revenue of $18 600 million, accounted for less than 4 per cent of California's GDP but used more than 75 per cent of water supply in the state. The central and coastal valleys grow more than 200 crops and are the predominant source in the USA for artichokes, processed tomatoes, almonds, apricots, dates, figs, grapes, kiwi fruit, nectarines, olives, pistachios and walnuts. The state has 30 million acres of farmland, of which 9.6 million are irrigated.[3] Employment in agriculture and related industries is a vital source of wellbeing to many rural communities, although these twin sectors are disfigured by extreme disparities in wealth. Of the state's 82 000 farms, less than 4 per cent produce more than 67 per cent of total revenues. Employment is seasonal, 80 per cent of the workforce are wage-labourers, seasonal migration is common and the majority of the workers are from Latin America (Gleick et al. 1995: 40; Villarejo & Runsten 1993).

According to the Californian Department of Water Resources, in 1990 Califor-

---

3. 1 acre = 0.405 hectares.

nia agriculture's "applied water demand", that is, the volume delivered to the farm gate, equalled 30.6 million acre-feet. (1 acre-foot of water equals $1182\,m^3$). In the same year, the **groundwater overdraft**, that is, the excess of abstraction over long-term recharge, was estimated to be 1.3 million acre-feet in more than 30 separate water basins. The situation is particularly unfavourable in the Tulare Lake, Central Coast and San Joaquin River hydrological regions. In round terms, in irrigated agriculture, 50 per cent of consumption is through furrow and flood irrigation, 35 per cent through sprinklers, and 15 per cent through drip and micro-sprinkler techniques (Gleick et al. 1995: 5, 42, 66; Sunding et al. 1994).

Insect pests and weeds are controlled by a petroleum-based chemical arsenal, which may be losing its potency. In addition, chemical fertilizers are used to promote yields. Excessive abstraction of groundwater is associated with worrying problems of soil and water pollution. In Los Angeles and Monterey counties, sea-water intrusion is occurring. Excessive abstraction also accelerates the movement of contaminants within aquifers and overpumping draws water-borne pesticides and nitrates towards wells pumping water for human consumption (Gleick et al.

Figure 7.1 California.

1995: 6–7, 45).

Furthermore, the routine chemical contamination of aquifers through agricultural drainage takes place, particularly in the San Joaquin Valley where the water table is close to the land surface. This necessitates:

the construction of drainage systems to keep groundwater tables from coming too close to the surface where salts can leach out of accumulated irrigation water. The drainage water is heavily salinized and in some areas contains concentrated levels of naturally occurring selenium and molybdenum (which) when concentrated in drain water cause problems for wildlife.

Even in the absence of overdrafting and shallow water tables, pollution from agricultural runoff is contaminating both surface and groundwater with nitrates and pesticides, such as the soil fumigant dibromochloropropane (Gleick et al. 1995: 45–6).

Table 7.1 presents data for 1990 on California's irrigated crop acreage. Crops are ranked by their revenue per acre-foot of water used. This water productivity variable is a close cousin of that used by the German Development Institute in their work on Jordan, as reviewed in Chapter 4. The GDI specification is to be preferred to that of the Pacific Institute, since it is based on value added, not gross sales. Water productivity in Table 7.1 varies from $130 per acre-foot for pasture, right up to $2273 per acre-foot for "other truck" products. Even the briefest glance at the table indicates that the scale distribution of agricultural land between crops is

Table 7.1 Californian irrigated crop acreage and revenue per acre-foot of water used in 1990.

Crop	Irrigated acreage ('000)	Revenue per acre-foot of water ($)*
Pasture	955	130
Alfalfa	1134	162
Rice	517	162
Other field	491	195
Corn	403	260
Cotton	1244	325
Sugar beet	216	325
Grain	988	357
Almonds and pistachios	510	747
Other deciduous	570	942
Vineyard	748	1201
Tomatoes	352	1234
Subtropical	419	1331
Other truck	1024	2273
Total	9571	n/a

Source: Gleick et al. (1995: table 11 and fig. 8).
* Data for 1988. Water is defined as the proportion of the total evapotranspiration provided by water applied through irrigation.

totally unrelated to their relative water efficiency. The crux of the quantitative water problem of California is that the distribution of water between crops is not positively correlated with water productivity. In 1990, in round terms, half of water used produced only 15 per cent of total crop revenue, whereas one-fifth produced 60 per cent (Gleick et al. 1995: fig. 10)

Having reviewed the *current* state of affairs in Californian agriculture, one now looks at the Pacific Institute's vision for 2020, beginning with its quantitative dimension. Gleick et al. present two scenarios, of which only the first is set out here (Gleick et al. 1995: 65–74). Scenario 1, entitled "Balanced groundwater", provides concrete estimates for the changes necessary to eliminate unsustainable groundwater use. It has five assumptions:

- no improvement in overall irrigation efficiency
- no improvements in crop yields
- irrigated alfalfa and pasture acreage within each hydrological region is reduced to the point where the amount of water saved equals the amount of groundwater overdraft projected by the Department of Water Resources for 2020 (i.e. 1.0 million acre-feet)
- in variant 1a, cropland freed by alfalfa and pasture reductions is left in a fallow state
- in variant 1b, cropland freed in this way is proportionately reallocated to all the other crops already grown in the region.

The results of Scenario 1 are given in Table 7.2. In both variants, the irrigated crop area falls, water consumption falls to eliminate the forecast groundwater overdraft, and crop revenue rises. In Scenario 2, where the additional assumption is made that rice and cotton acreage will be scaled back to 1960 levels, the results are generally more dramatic, but only variant 2b secures an increase in total crop revenue.

Table 7.2 Results of the Californian balanced groundwater scenario.

	1990	2020 1a	2020 1b
Crop area irrigated ('000 acres)	9571	8918	8998
Water consumed ('000 af)	21261	19137	19137
Crop revenue (million 1988 dollars)	12191	12340	12645

Source: Gleick et al. (1995: table 18).

In the "Balanced groundwater" scenario above, the first assumption is that there would be no improvement in irrigation efficiency. In fact, the Pacific Institute team do believe technology gains are possible, alongside improved information systems for water use on farms. Data are required on how much water is actually applied to and evaporated from crops and how much returned to groundwater, as a function of crop, irrigation method, climate and soil type. All groundwater use and quality should be monitored and managed by local groups for that purpose, guided by statewide standards. On-line data collection and dissemination networks should be developed to give farmers immediate meteorological and hydrological information (Gleick et al. 1995: 4, 17). As Gleick et al. wrote (1995: 82):

> On-farm irrigation water is useful to farmers only when that water goes to grow crops or to leach unwanted salts from the root zone. Excessive use of irrigation water leads to increased evaporation, unintended percolation to groundwater, and unnecessary runoff. Often, excess runoff carries with it agricultural chemicals, such as fertilizer nitrates. While increased irrigation efficiency can reduce water losses and protect and enhance water quality, improvements in efficiency can sometimes lead to lower water quality, or to reduced leaching of salts from soil. The decision about how to best manage irrigation water is thus a complex one . . .

Certainly, the irrigation technologies are available, including sprinkler and drip irrigation systems, agricultural water station networks, real-time moisture monitoring, and central computer controllers. Pinkham (1994) suggested reduced water use and improved runoff quality would follow wider application of laser levelling, surge valves, and **tailwater** retention ponds.

The reduction of distribution losses is another technological option. Substantial quantities of agricultural water can be lost somewhere between the headworks and the farmer, through seepage from unlined irrigation canals and through evaporation from aqueducts and reservoirs. The economic loss is that the headworks and network infrastructure are operating below their capacity, in terms of water deliverable to agriculture. The hydrological loss to groundwater stocks is only through evaporation, or when seepage is chemically contaminated, or where it runs to a saline sink.[4] In California, SCEA strongly indicates leakage reduction in many cases where water saved is used by urban water utilities. However, in the case of the All-American Canal on the Mexico/USA border, reduced seepage threatens the groundwater supplies of the Mexicali Valley, an unusual case of an international water conflict.

Next, one considers the sustainable vision for 2020 in respect of the quality of water and the use of water for nature conservation purposes. Much of this concerns farming practice. First, integrated pest management, spurred by a rethinking of fertilizer and pesticide use in the 1980s and 1990s, is strongly supported by the Pacific Institute Team. This leads to a reduction in chemicals application to the land and so reduces the pollution of water. Integrated pest management includes the use of cover crops, natural composts and mulches, and new disease-resistant crop varieties, and habitat provision for beneficial insects and birds. Secondly, permanent removal from irrigation is recommended for marginal lands susceptible to severe water quality problems in the southern regions, the Central Valley and the San Joaquin Valley. Thirdly, support is given to the strict control of agricultural drainage to protect groundwater in the state's vulnerable regions. Fourthly, flooding of

---

4. The economic evaluation of leakage losses is dealt with in Chapter 5.

rice fields by farmers is providing large quantities of food and outstanding habitat for migratory wildfowl and shorebirds, and the bird population reciprocates with its own brand of natural fertilizers (Gleick et al. 1995: 4–7, 51–2).

These quantitative and qualitative practices may be welcome, but how can they be achieved? The *non-economic* instruments of policy proposed in *California water 2020* include a new state and local groundwater management system; the protection of special flora and fauna habitats from urbanization through land-use planning legislation; technical assistance at county level, for example in habitat corridor programmes along berms, ditch banks and canals; information dissemination and water audits; and new modules in agricultural training courses.

The recommended *economic* instruments of policy include fiscal subsidies for both conservation initiatives and the fallowing of poor-quality agricultural land. Transferable abstraction rights are supported, but with a distinctly cool voice, and concern is expressed about potentially adverse impacts on rural communities and the environment; the text is thin here (Gleick et al. 1995: 86, 105).

However, the Pacific Institute team does give its strong backing for changes in pricing policy. In part, this concerns federal and state subsidies for crops with low water productivity. The clear and common-sense presumption is that this will lead farmers voluntarily to withdraw from crops such as irrigated pasture and alfalfa. Zilberman et al. (1993) showed that some 40 per cent of Californian crop acreage receives some form of fiscal subsidy (Gleick et al. 1995: 5, 17, 86–8, 104). In addition, the team fully supports price increases for water supplies. As Gleick et al. say (p. 82): "Without exception, experts on agricultural irrigation efficiency contacted for this report cite the most important incentive to increase efficient water use is to raise the price of water, which for farmers, is almost always subsidized." The rate basis, says the report, should reflect both the full capital costs of water infrastructures and the cost of the water utilities' programmes for conservation and efficiency. Increasing block tariffs are favoured. Anticipated consumption reduction is seen as the result of crop switching and of the more careful use of irrigation water on existing crops. The team expresses no view on the shape or elasticities of the effective demand function for irrigation water, but evidence from San Joaquin drainage areas is reviewed and is supportive of the beneficial effects of increasing block tariffs with respect to water use, the volume of drainage, and salt discharges.

7.7 Final remarks

This chapter began by arguing that the promiscuous use of the word "sustainable" to give cachet to virtually any policy development approved by a speaker or author should not blind us to the vital part the concept of sustainability will play in the future of water resource planning.

The roots of the concern with sustainability are traced back to the development of capitalism on a world scale and to that system's prodigious demands on the natural world as resource supplier, waste sink and life-support servicer. This has brought with it, over a period of little more than two centuries, widespread destruction of species and habitats, the depletion of non-renewable resources, the degradation of natural and built environments and the disablement of life-support services. I refer to these four processes collectively as inquination.

Since the mid-1960s, inquination has provided the context for the powerful growth of environmentalism as a social and political force. Moreover, the influence of environmentalist ideas on government policy is strongly buttressed by a growing recognition among political and industrial elites that global environmental change poses a threat to the long-term viability of the industrial economies.

The argument continues by proposing the adoption of the term "the sustainable society", that is, a society in which, for the indefinite future, human communities sustain and regenerate the species and habitats of the natural world, sustain and rehabilitate the quality of natural and built environments, sustain the global ecosystem's power to provide life-support services, and sustain and transform society's economic capacity to meet the material and cultural needs of its people. The mother concept in turn can give real meaning to derivative, sector-specific concepts such as sustainable development, sustainable cities and sustainable transport.

The quest for the sustainable society requires a broad spectrum of policy innovations. Within that spectrum, water resource management is a key activity. In this context, a fundamental rule providing the basis for action is proposed. The sustainability principle states that water resource management for a sustainable society should seek protection of the hydrocycle; purification of water from domestic, agricultural and industrial effluents; conservation of species and natural habitats in all their fresh- and coastal water environments; husbandry of water in its production and use; the supply of fresh water sufficient to meet the biological, cultural and economic needs of the human population; and, finally, the drainage of stormwater alongside the protection of rural and urban communities against floods.

The first case study was directed to the valuable policy work of the UK's Royal Society for the Protection of Birds, in particular its proposals for the protection of the hydrocycle. The principal weakness of the RSPB's strategy document is that it excludes all questions of water quality. The second case study, from California, is a paradigm of sustainability research, because of the breadth of its sectoral interest, its inclusion of both quantitative and qualitative dimensions, the depth of its analysis, and the persuasiveness of its policy recommendations.

CHAPTER EIGHT
The environmental costs and benefits of water projects

8.1 Introduction

Chapter 7 sought to place the environmental concerns of hydroeconomics within the broad context of debates on the sustainability of human society. Now one moves back again to the narrower scope of project analysis, with the purpose of asking how the economist should deal with environmental impacts of water projects. Such impacts may be negative, as in the case of a dam drowning a beautiful valley or displacing tens of thousands of rural citizens. Impacts may also be positive, as in the case of the restoration of a wetland, or reversing the pollution of a lagoon with a new wastewater treatment plant, or reducing infant mortality by raising the quality of the freshwater supply. Note that the concept of "environmental impact" used here embraces both the natural world and human communities.

8.2 Dual analysis

One approach to incorporating environmental change into project evaluation can be called "dual analysis". As one might imagine, this requires an appraisal combining two separate procedures. The *first* of these procedures is the calculation of the net present value of an independent project or the net benefit–investment ratios of interdependent projects. As we have seen, SCBA's focus of interest is the distribution over time of net changes in gross domestic product as a result of proceeding with a project in comparison with not proceeding with it.

All projects have environmental impacts in the sense that, in their gestation or their working lives, they induce weak or strong changes in defined ecosystems. In many cases these modifications bring about GDP effects. For example, the liming of a river to ameliorate acidification can increase the salmon population, with a consequential rise in fishing licence fees paid to the owners of the river's angling rights (Merrett 1992). From an economic point of view, the recreation services supplied by the river improve and this is reflected in the additional economic rent earned by the river's custodians, a differential recorded in gross domestic product growth. Where a project has environmental impacts which clearly raise (or lower) GDP, then these with-project effects should be estimated and handled quantitatively in the manner described in Chapter 5.

However, most of the environmental impacts of projects do not have a clear GDP effect, nor do projects' redistributions of income and of life-chances within human populations. In this situation many economists regard these environmental and distributive outcomes as priceless, above price, beyond price. Thus, it can be argued that effects beyond price should be handled separately and differently from effects that are priced and which appear in the SCBA calculation (Merrett 1994). Environmental and distributional outcomes are often inseparable, as is illustrated by dam-building displacement of human populations or falls in morbidity attributable to wastewater programmes. This brings us to the *second* procedure which, alongside SCBA, makes up what is here called dual analysis, and that is environmental impact assessment (EIA; see Glasson et al. 1994). A typical environmental impact statement is complex, but its key features are:

- it is carried out by professionals with a training in the environmental sciences
- it is multidimensional in the environmental issues it addresses in respect of water, soil, air, ecosystems, species and human communities
- it contains an appraisal of the no-project situation as well as the with-project scenario
- it combines textual appraisal as well as the quantitative measure of, for example, changes in biological oxygen demand of a river or the scale of schistosomiasis in local villages
- it states the technically feasible mitigating measures that could be taken to reduce the scale of negative environmental impacts.

Note that environmental effects beyond price may impact on the persons and institutions directly involved in the project's gestation, production processes and the use of its outputs. Or these effects may be external to the project. It is important to understand that the distinction between internal and external effects is separate from the distinction between GDP effects and effects beyond price.

With social cost–benefit analysis and environmental impact assessment in place, the distributional outcomes for specific population groups can be identified. Such groups may be classified by gender, income-category, class, ethnicity, district or region, and so on. Outcomes will be environmental (see above) or economic, such as job opportunities, consumption shifts, income change and training initiatives. Distributional studies are here considered as an integral part of dual analysis.

Dual analysis requires that, for independent projects, the net present value and the EIA should be brought together, compared and contrasted, and the decision-makers must at this point use their human wisdom to make a final judgement on whether or not the project should be financed. Where projects are interdependent, for technical or funding reasons, the EIA should be used to reorder the project rankings initially based on the net benefit–investment ratio.

So, dual analysis – combining the net present value and the EIA for each project, whether independent or interdependent – should be the key to project selection. But notice that no simple or quantifiable procedure exists for the resolution of dual analysis choices. This is simply because net present value and the EIA use different languages and, even in the case where both engage in quantitative estimation, these quantities are not commensurable.

In the limiting case, effects beyond price may be of such great importance in the rationale for a project in the first place that the measured value of the incremental real output is a secondary issue. In this case, dual analysis proceeds by combining the present value of the incremental real costs with the EIA, for a go/no-go decision. Where interdependent projects are broadly matched in their effects beyond price, that with the least present cost is to be preferred, of course. Here one returns to social cost-effectiveness analysis.

For the river basin and water infrastructure programme (or the national economy as a whole), where many projects are approved in which the "outputs" are not marketed and so are financed from the public purse, the public finance implications of such projects in aggregate must be constantly reviewed. More generally I concur with the view of Little & Mirrlees (1991: 376) that ". . . the fiscal effect of all projects should be estimated and included in appraisal reports".

The advantage of dual analysis is that it does not marginalize effects beyond price, but allows them to be used fully in the approval or rejection of independent projects as well as to re-rank all interdependent projects. At the same time the SCBA approach can be used to show the shift in the aggregate net present value as a result of such re-ranking. This re-ranking, with its inevitable cut in aggregate net present value (see Ch. 5) can be regarded as the net present value opportunity cost of employing the effects beyond price criteria. Note that the internal rate of return cannot do this since individual internal rates of return cannot be summed.

8.3 Monetizing the environmental impacts

But dual analysis is only one approach to incorporating environmental effects into project evaluation. The second approach is by means of **extended social cost–benefit analysis**. Since SCBA is entirely conducted in terms of money values, the US dollar for example, this extension of the conventional analysis requires the monetization of environmental impacts with no clearly measurable GDP effect, referred to earlier as "effects beyond price". This is done by assigning a dollar value to each such environmental outcome. The strength of this second approach is that all a project's GDP and its GDP-neutral environmental outcomes can be aggregated into a single value. A second strength proposed by the proponents of monetization is that placing a dollar sign against environmental outcomes has the political effect of making the defence of ecosystems and threatened communities more persuasive (Bateman 1995: 48).

Before looking at the various techniques of monetization that have been proposed, a few preliminary clarifications will be useful. First, monetization texts often call for the use of opportunity-cost prices in extended SCBA. This issue has already been covered in Chapter 5 under the discussion of shadow pricing. Secondly, the literature usually reviews dose–response approaches. This is a valuation method in which one begins by observing the marginal environmental impact (the "response") of successively larger-scale or more intensive project activities (the "dose"). Such impacts are then valued, as for example in the case of air-pollution damage to buildings. In practice, the estimation of the dose–response relation can be enormously difficult, because of thresholds, synergetic effects and lagged impacts (van der Straaten 1994). The dose–response technique can be used either for calculating the GDP effects of environmental impacts or with GDP-neutral environmental outcomes. Hereafter, the focus is on the attempts to monetize the GDP-neutral impacts of projects, that is, environmental impacts with no clearly measurable GDP effect.

The monetizers frequently begin their exposition with a treatment of value theory (Pearce et al. 1989: ch. 3; Turner & Postle 1994: 5–11). It is pointed out that our motivation in seeking environmental benefits or avoiding environmental costs is many-layered. If one considers the conservation of a wildlife habitat, such as the RSPB's magnificent Minsmere reserve in Suffolk (England), for many the value is clearly that of using it for a day's birdwatching, with a prospect of seeing avocets, bearded tits, bitterns and shelducks. Even if one is not a present user, one may well value the option of using the reserve in the future. At the same time, the thought that one's own generation creates a place of beauty and biological diversity, which will be available for generations to come, brings a warm glow, now, over and above those future uses. Finally, one may value the existence of wood, seashore, scrape, pool and reedbed, now and for a thousand years, irrespective of the pleasure oneself or other persons, today or tomorrow, derive from it.

These commonplace attitudes are sometimes labelled by economists with the titles, respectively, of primary use, option value, bequest value and existence value, and environmental economics is construed, reassuringly, as merely an extension of "conventional economic theory" (Bateman 1995: 61). Primary-use environmental outcomes of projects can often be measured in GDP terms. Option value, bequest value and existence value outcomes are always GDP-neutral.

The monetization of GDP-neutral environmental impact takes two entirely separate courses, called here variant costing and pseudo-market pricing. In the case of variant costing, the starting point is the EIA. On its basis it may well be possible to specify project variations by which negative environmental effects can be mitigated or positive effects reinforced. Usually, mitigation costing is the issue and this can be achieved by measures such as project redefinition, as in the case of shifting from a proposal to dispose of wastewater by coastal dumping to one for wastewater treatment. Another example, from Wales, is the Cardiff Bay barrage's threat to mudflats used by wintering wildfowl. This led to the Development Corporation committing itself to the creation of alternative feeding grounds elsewhere. Such variations can be costed and reviewed within the standard procedures for project selection (DoE 1991: 24).

In the case of pseudo-market pricing, one begins by recognizing that no formal markets exist in which projects' environmental impacts are priced. So, indirect means are found to estimate what people would be willing to pay, if such markets *did* exist. Here, three techniques deserve consideration: the contingent valuation method, the travel cost method, and the hedonic pricing method.

To recapitulate, whereas variant costing concerns itself largely with estimated expenditures on measures for the mitigation of negative environmental effects, pseudo-market pricing investigates the prices that might be generated by the values people place on the environment. Variant costing is a supply-side approach and pseudo-market pricing a demand-side approach.

Contingent valuation has wide application in the field of hydroeconomics. If we use the concepts of primary user value, option value, bequest value and existence value, as referred to above, the contingent valuation method can be applied to all four and, in principle, it has relevance to every one of the seven key subject areas of the economics of water laid out in Chapter 1.

The method is implemented by interview surveys of an appropriate sample of the population affected or likely to be affected by a specific project, plan or policy. Information is sought on the maximum amount a person would be willing to pay to enjoy an environmental benefit or to avoid a disbenefit. An anchor price is set in the questionnaire and, provided the respondent is willing to pay more than this, the price is successively raised until the maximum is arrived at.

An alternative is to set an anchor price the consumer-citizen would be willing to accept in compensation for the loss of an environmental benefit or the inception of a disbenefit and, provided the respondent is willing to accept less than that, the price is successively reduced until the minimum is arrived at.

The maximum payment or minimum compensation associated with these expressed preferences is then multiplied by the size of the total population affected, to give the aggregate contingent value. This contingent value can then be entered into the net present value calculation in social cost–benefit analysis extended to incorporate GDP-neutral outcomes.

The contingent valuation method has been used, for example, in a study of water quality improvement for the Monongahela River in the USA, with respect to recreational fishing, boating and swimming, and in a study of the benefits of coastal defence measures that would stop the erosion of recreational beaches (Desvousges et al. 1987, Green et al. 1990, Penning-Rowsell et al. 1992).

Next the *travel cost method* is considered. Unlike the contingent valuation method, the travel cost method has only a narrow application to hydroeconomics, essentially restricted to studies of primary user visits to nature reserves and the like. A survey is carried out of trips to a waterfall, for example, and the cost of the trip is estimated, aggregating entrance fee (if any), the money costs expended on the journey and the value of the time spent making the journey. This aggregate cost is taken as the price of the revealed preference to make the visit. The survey also records the number of visits made by the respondent in the year. This price and quantity for one respondent is pooled with the rest of the survey sample to give an effective demand function for the recreational facility. After converting the demand data to reflect the position of the entire population of visitors in the year, the area under the demand curve is calculated for the known total of visits and is used as the estimated recreational value of the waterfall per year. An example of the application of the travel cost method is to the recreational use of inland water-ways (Willis & Garrod 1990).

Like the travel cost method, the *hedonic pricing method* is also narrowly limited in its hydroeconomic application, relating only to the availability of water resources to local residents. Data is collected on house prices in a given area and the value of variables that might explain house price differentials. The data are usu-ally cross-sectional, such as catchment with catchment, rather than time series. Multiple regression analysis is then used to estimate the price people are willing to pay, as part of the total price of the house, for each unit enjoyed of each of these independent variables. Where, for example, access to a local lake is identified as one such variable, the differential house price effect is calculable. Innumerable hedonic pricing method studies exist of house (and land) price differentials, but a water dimension is either absent from them or is a variable of slight importance.

8.4 A critique of environmental monetization

The preceding section provides a brief and uncritical exposition of monetization techniques. What follows here is an assessment of their methodological difficulties and, more generally, their conceptual validity.

Hedonic pricing
The hedonic pricing method's principal difficulty is that the determination of house prices is so complex. In general we would expect the physical characteristics of the house to be the dominant set of factors, followed by the location of the dwelling with reference to travel-to-work and, thirdly, the qualities of the neighbourhood, such as the reputation of local schools, street crime, traffic noise and pollution (Merrett with Gray 1982: ch. 5). This complexity means that the survey costs of collecting information are high and problems exist in measuring many variables – such as school quality. The mutual correlation of causative factors can make their inde-pendent effects impossible to measure, and non-inclusion or poor scaling of vari-ables will give low explanatory power to the multiple regression equation, with possible bias in the parameter values of those independent variables that are meas-ured accurately.

Moreover, as Bateman points out, having solved the hedonic price equation, one has to derive from it, in a second stage, the demand curve for the specific attribute of interest, and with water resource availability there may be no quantitative scale of use – just to have or not to have (Bateman 1995: 56–7). All this, alongside parameter values that are positive when common sense suggests they should be negative, or vice versa, puzzles in the choice of the mathematical form of the multiple regression equation, and the dangers of pooling data from areas with different house price determining structures, suggests that the hedonic pricing method will probably never be of use in the environmental appraisal of water projects.

The travel cost method

The travel cost method's methodological problems have been well reviewed by Pearce et al. (1989: 87–90). The monetization procedure is applied to visits to an existing site. Since *ex ante* project evaluation of environmental benefits is, by definition, being carried out for a new site (or an improvement of the existing site), it is difficult to know how best to apply past results to a qualitatively new situation. Other methodological enigmas are how to cost a visit that is not a one-off trip but part of a wider tourist menu; how mathematically to represent the demand equation for trips, when other variables such as car ownership and trip substitutes can be powerful determinants of the number of visits per year per family; and how to allow for visits to a new (or improved) project by households excluded by travel costs from the surveyed site. Moreover, serious problems are encountered in valuing the time allocated to a visit since this cannot be seen as a simple trade-off with working time and, in any case, the value of working time is itself difficult to measure. At the other end of the scale, a travel cost method study of canal-based recreation was weakened because travel costs were negligible (Turner & Postle 1994: 21).

Contingent valuation

The contingent valuation method is strikingly different from both the hedonic pricing and the travel cost methods, in that its applicability to the environmental benefits and disbenefits of water projects is much wider. The central problem encountered by the technique, in my view, is the ontological nature of the estimates it generates. This requires further explanation.

In market economies – for all practical purposes the entire globe – the concept of a "market price" is extraordinarily concrete in what it denotes. It refers to the sum of money that users are willing to pay, are able to pay and which they actually do pay to producers or distributors per unit of a good or service. Simultaneously, it is a sum of money per unit product which producers and distributors are willing to accept and do accept in exchange for supplying each unit to users. Moreover, market prices – which frequently change from one time period to the next – in most cases bring a rough and ready correspondence in the quantities produced and purchased in each year, so that stocks of goods neither rise rapidly nor fall rapidly.

The concept of a "contingent price" is absolutely distinct. It refers not to a market transaction, either observed or predicted, but to a hypothetical payment which the respondent knows will not, in fact, be made. It has a fundamentally counterfactual quality.

The concept of a market price, in the real world, denotes money changing hands. The concept of contingent valuation, again in the real world, denotes a response to a questionnaire. They are quite different categories in terms of the real social processes that they denote. Given this disparity, to aggregate them commits what Gilbert Ryle, an Oxford philosopher, called "a category mistake" (Ryle 1949). It makes as much sense, in the extended SCBA of a dam project, to subtract the contingent value of the reduced population of Sumatran rhinoceros from the output value of hydroelectric power, as it does to subtract three buttercups from five watts or to add an apple to a logarithm.

So, the only usefulness of the contingent valuation method can be as a free-standing technique for assessing the strength of feeling people have about the environment. Even here, difficulties exist. Survey response rates can be low. "Free-rider" bias exists, that is, responses may be influenced by some people's belief that they will in any case enjoy free access to the project. Valuation can be contingent on the payment vehicle chosen. The measures of willingness to pay to avoid an environmental loss, and the willingness to accept compensation for the same loss, may differ by three orders of magnitude. Both are strongly influenced by the anchor price chosen in beginning the search for the desired maximum/minimum. The information presented to respondents can heavily influence their response. As with the travel cost method, people find it more difficult to price the differential effects introduced by a potential project rather than a known environmental resource. This was apparent in a study of the economic benefits of reduced acid deposition on aquatic ecosystems (Turner & Postle 1994: 19; ECOTEC 1993). In another willingness-to-pay study of the recreational use of a riverside park, "Nearly half of the respondents . . . could not or would not value their enjoyment . . . with most considering that they could not value such things in money terms" (Turner & Postle 1994: 26).

Variant costing
The final monetization technique for review here is variant costing. As we have seen, this does not attempt to establish a pseudo-market but sets out the estimated costs of project variations mitigating environmental disbenefits or strengthening benefits. Variant costing finds its place within dual analysis and is assessed by the increase in the range of options it offers, alongside the additional expenditure required to cost them.

8.5 Case study: saline flooding in Norfolk

The case study for this chapter is drawn from the work of Ian Bateman, Ian Langford and Andreas Graham of CSERGE, the Centre for Social and Economic Research on the Global Environment (Bateman et al. 1995).[1]

Bateman and his colleagues introduce the substantive issue in the following way:

The Norfolk Broads is a site of recognized national and international wildlife importance. Recently accorded National Park status, most of the area is a designated Environmentally Sensitive Area, and contains 24 Sites of Special Scientific Interest and two sites notified under the international RAMSAR convention. The character of the low lying landscape of the Broads is dependent on 210 km of river embankments for protection from saline tidal water. However, these flood defences are increasingly at risk from failure, both because of their considerable age and continuing erosion from passing river traffic and, more fundamentally, because of settlement and sinkage of the adjacent and underpinning marshes. Thus, the standard of flood protection afforded by these defences is decreasing over time. The consequences of increased saline flooding upon affected fauna and flora, recreation, agriculture, property and infrastructure are likely to be significant.

The project in this case, then, is a set of schemes to alleviate this flood risk.

The survey reviewed here is not one of site visitors to the Broads (Fig. 8.1) but a mail survey of persons not on site. These are referred to as non-users, although it is recognized that, as a group, the values they hold for Broadland embrace primary use, option, bequest and existence motivations. The research uses the willingness-to-pay technique and the sampling strategy seeks to determine whether willingness to pay declines with distance from the Broads and, also, whether a willingness to pay gradient exists across variables such as social class and income. This was done to permit "a defensible aggregation across Great Britain" of the willingness to pay results.

The mailing to each person sampled gives basic information on the Broads and the likely consequences of not taking initiatives to alleviate saline flooding. The text contains a colour map detailing the extent and salient features of the area, defining the likely limits of flooding expected in the absence of further investment in river defences.

There is no anchor price or bidding sequence of the kind that interview surveys permit in estimating willingness to pay and willingness to accept compensation. However, a budget question, asking the amount the household spends annually on countryside recreation and preservation, may have acted as an anchor (Bateman et al. 1995: 24).

The payment vehicle to implement the willingness to pay is posited to be a national increase in personal taxation, so that respondents would have been aware of the counterfactual nature of the method. Bateman et al. seem to believe that their respondents would be "genuinely prepared to pay the sums stated". But no evidence is provided to verify this belief.

---

1. CSERGE is a research institute attached to University College London and the University of East Anglia. The Centre is the strongest location in the UK for economic research into the environment. The work of its staff, such as David Pearce, Kerry Turner and Ian Bateman, has been drawn on extensively in the writing of this chapter.

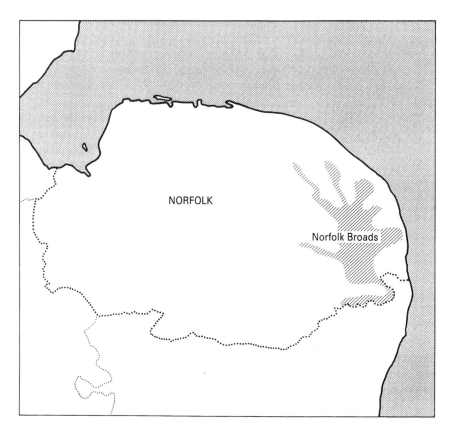

Figure 8.1   The Norfolk Broads

The survey's results will now be reviewed. The 34 per cent response rate is disappointing: 344 responses were received from a random mail-out – based on postcode registers – of 1002 questionnaires. Non-usable responses (34) reduced the database to 310 entries. Data analysis by the research team also indicates that response bias exists, with a strong overrepresentation of higher income, relevant user and countryside membership groups. Mean expenditure on countryside recreation and preservation was £265, but the spread of estimates was huge, with a standard deviation of £280.

To the question of whether or not the respondent was willing in principle to incur extra personal taxes to pay for flood defence in Broadland, 54 per cent answered positively. The research authors write: "Most importantly very few respondents appear to have objected to the fundamental principle of valuing the preservation of Broadland" (Bateman et al. 1995: 17). This ignores the probability that fundamentalist objectors are likely to have been concentrated in the 66 per cent of the households surveyed who did not respond to the questionnaire. After all, why spend time filling in a willingness-to-pay questionnaire if you believe it is fundamentally misconceived? In any case, the upshot was that only 17 per cent of the total sample positively expressed a willingness to pay extra taxes for flood alleviation.

Mean willingness to pay of the 310 usable responses is £21.75 per year, counting as zero the willingness-to-pay value for the 46 per cent not willing to pay extra taxes. As expected, the distance decay hypothesis is confirmed: willingness to pay falls markedly with the respondent's distance from the Broads. However, there was no statistically significant correlation between socio-economic status and willingness to pay. The mean willingness to pay expressed as a lump-sum, rather than on an annual basis, was £50.86.

The research results also show that the willingness to pay of the off-site survey of £21.75 was less than one-third of an on-site survey of Broadland users, a logical relationship. Moreover, reviewing six user and non-user surveys of natural recreation facilities, willingness to pay is greater the more substantial the project, and smaller the more alternative forms of recreation are available. Once again, this is a logical relationship.

Finally, one needs to examine how ready are the CSERGE researchers to extend SCBA, by using these individual respondent willingness-to-pay calculations to arrive at an aggregate value stream, as is done in Chapter 5 with SCBA's use of the net benefit flow in GDP terms. Here their initial position appears confident enough, for they state that the "sum" of direct use value (e.g. fishing), indirect use value (e.g. recreation), option use value (e.g. future personal recreation), bequest value (e.g. future generations' recreation) and existence value (e.g. preserving wildlife habitats) "tells us about how much humans value a particular resource. Such a definition clearly expands upon the market-orientated approaches of many simple financial appraisals of projects" (Bateman et al. 1995: 2). In this vein, the sampling strategy seeks to allow a defensible aggregation of the willingness to pay responses across Great Britain.

In a wonderful moment of spurious precision, they suggest that a conservative adaptation of the results to allow for the survey's response bias gives an enormous overall valuation of £109.8 million pounds in Britain as a whole for the environmental benefits of flood alleviation. They say that the sample values should, arguably, be considered by decision-makers. They show, too, that conventional SCBA gives a benefit:cost ratio of only 0.94, but that adding in willingness to pay, from on-site and off-site respondents, increases the ratio to more than 4. But by the end of the study, Bateman et al. are exhibiting acute signs of schizophrenia. They note the view of critics of the contingent valuation method, who argue that the controversy surrounding the method is so great that the proposition "some number is better than no number" is invalid and dangerous (Young & Allen 1986, Diamond & Hausman 1994). The Bateman et al. response to such trenchant criticism of the contingent valuation method is that they do not reject such a viewpoint. Moreover, they had already suggested that willingness-to-pay responses can be considered as "at best very approximate estimates of values and at worst spurious guesses primarily motivated by a subconscious desire to support a 'good cause' irrespective of the particular good under evaluation" (Andreoni 1990). Here, the interview question on the respondent's annual budget on countryside recreation and preservation acts as a critical anchor for the willingness-to-pay response. One should note that such an anchor would limit the statistical variance of the willingness to pay and therefore raise the explanatory power of the multiple regression equations used to test contingent valuation method hypotheses.

In summary, CSERGE's contingent valuation method study of saline flooding in Norfolk can be considered as an excellent example of the method's use in the field of hydroeconomics. However, it does nothing to remove doubts over the counter-factual nature of the technique and it exhibits serious ambiguities by the research team on the method's validity.

8.6 Final remarks

This chapter has sought to build on the Chapter 5 treatment of social cost–benefit analysis for water projects, by considering how projects' environmental impacts should be handled.

The first approach reviewed is that of dual analysis, combining two procedures: SCBA and EIA. Where environmental impacts contribute to a raising or lowering of GDP, this dimension of the impact should be incorporated in the SCBA. However, most of the environmental impacts of projects do not have a clear GDP effect. These outcomes can be referred to as effects beyond price or, alternatively, as GDP-neutral impacts. In dual analysis they are handled separately and differently from effects that are priced and which appear in the SCBA calculation. This procedure is the environmental impact assessment. Both qualitatively and quantitatively, the EIA is essentially multidimensional.

Project decision-making is then based on both these distinct methodologies and it incorporates the distribution of economic and environmental outcomes. Since the results of one cannot be added to the other, a choice between two competing projects cannot be made on the basis of which scores highest in a quantitative sense. Instead, the selection between alternatives requires our human wisdom – a scarce commodity – because the impacts of projects are recognized as non-commensurable.

The second approach reviewed is extended SCBA. Here, GDP-neutral impacts are priced and the summed net benefits from such environmental effects are added into the traditional SCBA net benefit flow to produce a revised net present value. Extended SCBA is based on value theories specifically adopted for environmental research. As with SCBA, it never considers the distributional outcomes of projects. A widely used categorization for environmental evaluation is that of primary use value, option value, bequest value and existence value. The monetization of environmental impact uses four techniques: variant costing, hedonic prices, the travel cost method and contingent valuation.

In fact, the variant costing method is best considered as a valuable weapon within the armoury of dual analysis and one which has wide applicability. The EIA is used to specify project variations by which negative environmental effects can be mitigated or positive effects reinforced. Each variation both changes the measured costs of a project and at the same time changes the nature of its environmental effects, so each variation constitutes a distinct project option.

The hedonic pricing method, like the travel cost method and the contingent valuation method, is a pseudo-market technique in which, because of the absence of a formal market in environmental outcomes, indirect means are found to estimate what people would be willing to pay to enjoy environmental benefits or would be willing to accept as compensation for environmental benefits forgone. The hedonic pricing method is costly, faces serious technical difficulties and has enjoyed little application in the analysis of water projects. Within the discipline of hydroeconomics, at the very least, it is a waste of time and resources.

The travel cost method, similarly, has only a narrow application in the field of water services, being largely restricted to the analysis of environmental recreation. It faces serious technical difficulties and, like hedonic pricing, is a waste of time and resources.

Contingent valuation, in contrast, is used to deal with all four of the human value categories of environmental economics and can be applied to every subject area of hydroeconomics. Here the pseudo-market is created through personal interviews or mailed questionnaires aimed at eliciting willingness to pay or willingness to accept compensation. It faces some technical difficulties, as the case study showed. However, the principal objection is that adding net benefits base on contingent valuation to a traditional SCBA's net benefit stream commits a logical error. For the contingent valuation method – as the word "contingent" indicates in this case – is a counterfactual method which produces aggregate dollar values of a completely distinct category from those of SCBA's GDP differentials. The contingent valuation method is, therefore, based on a category mistake and for that reason should not be pursued.

In conclusion, the proper study of projects in respect of their GDP and environmental outcomes is by means of dual analysis. Moreover, by documenting the distributional outcomes of the economic and environmental effects of projects, this enormously important dimension to project evaluation is also incorporated.

CHAPTER NINE
Political economy and water resource policy

9.1 Hydroeconomics: content and approach

In this final chapter, an attempt is made to weave together the principal arguments of the Introduction and Chapters 2 to 8. Economics has been defined as the study of the nature, processes and social relations governing resource allocation, that is, the production, distribution and exchange of the requisites of human life (EAEPE 1991b: 1). Hydroeconomics is the economics of water resources. Water is a natural resource in a hydrological sense and, biologically, is one of the fundamental requirements of all life on Earth. Water is also an economic resource, and always has been so for the human race, because labour and the means of production are necessary to abstract fresh water and bring it to its point of use. Water's status as an economic good also derives from the fact that (unlike sunlight and the air) rivers, lakes, estuaries and coastal waters can all be appropriated into the ownership of public or private bodies. Water's status as an economic resource, rather than as a free gift of nature, has been reinforced in the past 15 years because of a global shift to the privatization of public sector infrastructures, because laws to protect the quality of water have imposed much higher treatment costs, and because the competition between users for scarce water has become more widespread and intense.

The primary orientation of the economics of water resources is to water which is abstracted, stored and distributed by human labour, to the use of that water, and to the disposal of wastewater. So, the seven subject areas of hydroeconomics are:
- nature conservation for rivers, lakes, wetlands, estuaries and coastal waters
- land drainage
- flood control and coastal defence
- dam projects
- the supply of fresh water
- the use of water by households, agriculture, and industry
- the treatment of wastewater and its disposal.

Because economic analysis divides into distinct paradigms, an explicit choice is demanded between three competing schools of political economy in order to construct the economic analysis of water resources. These paradigms are evolutionary political economy, Marxian economics and neoclassical economics. In the writing of this text, I have adopted the first of these. It is characterized by its evolutionary approach, its institutionalism, its epistemology, its rootedness in ethical values, its open-mindedness and its concern for the interdependence of economy and environment. A subsidiary goal of this book is to contribute to the development of this paradigm by applying its perspectives to a specific economic sector.[1]

9.2 The hydrosocial cycle

Evolutionary political economy focuses on the substance of the fundamental processes involved in providing necessary goods and services for humankind. For hydroeconomics, a key concept here is the hydrosocial cycle. This comprises water abstraction, its storage, freshwater treatment, its distribution to users, their consumption of the product, the collection of wastewater, wastewater treatment, and its disposal into the oceans or, through recycling, into surface and groundwater sources. Fresh water, excluding that located in saline inland seas or locked in ice caps and glaciers, makes up only 0.6 per cent of the total of water on planet Earth.

In respect of the freshwater supply, an important debate is whether treatment should be to uniformly high standards or whether consumers with relatively low quality requirements, such as agriculture and certain industries, in contrast to households, should receive water treated to lower standards and consequently at less cost.[2]

A second issue concerns the conceptual treatment of leakages from the freshwater distribution system. Traditionally, this has been regarded as a demand-side matter. However, in this book it is argued that leakages are a supply-side phenomenon, akin to the loss of consumer goods, such as refined sugar, between the factories where they are produced and household purchase in the corner shop or at the supermarket. If this argument is accepted, then for water resource planning purposes, the presentation of water flows in a catchment's hydrosocial cycle, or at the level of a region or a whole country, can usefully take the general form illustrated in Table 2.1. In this water balance statement, the net supply total is equal to the use total. The economic losses attributable to leakage are the costs of abstracting, storing, treating and distributing that water prior to its loss. At the same time, the costs of remedying such losses should be recognized in any programme for their reduction. The environmental losses of leakage derive from lower river flows, reduced volume of water in other surface sources, and a fall in the water table.

---

1. For an economic philosopher, an interesting, paradigmatic case study would be to compare and contrast this book with the neoclassical text by Spulber & Sabbaghi (1994).
2. Note that, for convenience, the term "industry" is used throughout this volume to refer to all consumers other than households and farmers.

The cyclical aspects of the hydrosocial cycle are threefold: internal re-use by a single household, firm or other institution; external re-use where wastewater is taken up by other consumers; and the recycling of treated (and untreated) wastewater back into freshwater sources. The first two reduce the abstraction demands on ground- and surface water sources, and the third resupplies those sources. Equation 2.5 proposes a new definition of the flow supply of water for abstraction in a catchment, suggesting it equals effective rainfall plus the total volume of recycled water corrected for the relative location of the points of abstraction and recycling.

The water balance statement and the recycling concept are developed into the measure of catchment self-sufficiency as well as (Eq. 2.9), the catchment stress indicator for a river basin, that is, net abstractions relative to effective rainfall.

9.3 The supply of fresh- and wastewater services

In this book, consideration of the economics of supply focuses on the fifth and seventh subject areas of hydroeconomics, that is, fresh- and wastewater services. The real resource costs of supply fall into the two categories of headworks costs, such as a sewage treatment plant, and network costs, such as drains. Real resources used are extraordinarily varied, including fresh water itself, land, the buildings and capital equipment of headworks, the network's pipes, sewers and irrigation channels, electric power, materials and human labour.

Supply-side analysis concerns itself particularly with cost per unit of output for different scales of production, different technologies and different qualities of fresh- and wastewater treatment. To do this the monetary cost of production and distribution is classified into prime costs, overhead costs, and their joint total. Product price or, alternatively, the fixed charge received by a utility per unit output, can then be seen as the source of a complex series of pay-offs per unit output. These pay-offs are illustrated in Figure 3.2, and include wages and salaries, power consumption, loan interest, amortization, shareholders' dividends and retained profits.

Cost functions are used to present prime, overhead and total cost per unit output, for example at each scale of output. Here the Marshallian distinction between the short term and the long term is vital. With the short-term function, average total cost falls markedly with higher output levels as staffing costs and overheads are spread over greater volumes of output. This is a hyperbolic function. For any given unit price, the higher the level of capacity utilization, the greater the gross margin and the fuller can be the flow of funds for net amortization and retained profits, which are the internal financial sources for the accumulation of capital. Low product price, high unit costs and government taxation of net profit all inhibit the possibilities of a catchment monopoly upgrading its facilities, such as in reducing rationing of local consumers or in raising treatment quality. This has been a problem particularly faced by state-owned enterprises.

The cost function used in long-term analysis once again places average total cost centre-stage, but in this case considers it for each of a set of values of incremental fresh- and wastewater services capacity, planned as "a single leap" from the current situation with all its historical, spatial and environmental particularities. It is suggested here that the general shape of this *ex ante* function is quadratic (see Eq. 3.1). In its graphical form, the average total cost curve falls at low and medium levels of additional capacity. This is an expression of the indivisibility of infrastructural provision in respect of headworks, the distribution and collection network and a variety of staffing costs. Indivisibility is the economic basis of the so-called natural monopoly that exists in freshwater supply and wastewater treatment. But at high levels of incremental capacity, average total cost rises, particularly because surface water and groundwater at a greater distance and depth, or of lower quality, have to become the sources of supply.

The long-run average total cost curve must be clearly distinguished from costs involved in an evolutionary development of the water sector, for example, a first investment commitment followed by a second round of construction, as an urban area's demand expands over time. The durability, specialization, immobility, indivisibility and lumpiness of infrastructural provision provide a strong economic argument for urban land-use planning and demand management of urban growth.

9.4 Consumption, effective demand and the price of water

Our species, *Homo sapiens*, traces its ancestry back to amphibians, so it is hardly surprising that the consumption of water is a fundamental condition for human existence. Water is the basic solvent required for all the body's functions. In addition, households use water for a multiplicity of purposes such as cooking, washing and disposing of body wastes. So, the economic production of freshwater services is one of a very few economic activities directly necessary for the reproduction of the human race. The other predominant uses of fresh water are in irrigated agriculture, where water is a biological necessity for the growth of crops and the raising of livestock, and in industry – broadly defined – for cleaning, cooling, power generation and so forth. The proportionate use of water by households, agriculture and industry varies enormously over time and place.

The effective demand for water is the quantitative relationship deemed to exist, at a specific place and time, between the quantity of water purchased for consumption and its unit price. Effective demand is determined by the tastes and habits of the three consumer categories; the price, quality and availability of substitutes; and consumers' economic capacity to pay for water. The responsiveness of quantity purchased to price differences can be measured by the price elasticity of demand. I argue that, in general, the effective demand function for water is cubic, as in Equation 4.2. No evidence exists to test this hypothesis to the limits, but it is consistent with known market behaviour, unlike orthodox theories which hold that the effective demand function is linear, or convex to the origin, or of a constant elasticity.

Water utilities' financial costs in supplying freshwater services can be met either from a fixed charge or from water pricing. Where water is priced in a free market of capitalist firms, it is argued here that the intersection of a supply and a demand curve, as proposed in orthodox theory, is not an appropriate approach to understanding price determination. Nor, I believe, is analysis of the individual firm's decision-making usefully founded on the intersection of the marginal revenue and marginal cost curves, again of neoclassical theory.

The approach here is to use the estimated short-term average cost functions for the financial year ahead and to theorize that price is set as a mark-up of average prime cost at normal capacity utilization. This administered price is not necessarily market clearing. Moreover, it should be seen as neither a short-term nor as a long-term price. Its short-term characteristic is that it is derived from a short-term cost function. Its long-term characteristics are that this year's short-run is the end-point of a past evolutionary process; and that the size of the mark-up is set with future long-term market responses in view.

In contrast with theorizing on the basis of an unregulated free market in water services provision, we can consider the situation where water services are provided by a state company as a social service. Here, water prices or fixed charges are set to cover only prime costs at normal capacity. Real world institutions can be found somewhere along the spectrum between the two ideal types and they take many, many forms. The market approach is distinguished as follows: water utilities are private companies; there is some form of state regulation, weak or strong; there is no cross-subsidy between user groups; consumption is metered; and turnover is generated by average cost tariffs. The social service approach is distinguished as follows: water utilities are government-owned; there is no external regulator; households' consumption is neither priced nor metered; other users may be metered; users with political clout are cross-subsidized; supply-fix philosophies predominate; and fixed charges are levied to cover the water utilities' prime costs.

In the case of the social service ideal type, the absence of pricing exists because of government's general political philosophy, because of the supposed distributional consequences of pricing and because of the power of specific consumer groups on government policy. In situations intermediate between the social and market extremes, the absence of pricing also exists where private water utilities, having a fundamental goal of capital accumulation, see pricing – as an alternative to a fixed charge – as a constraint on the quantitative expansion of their product sales.

In respect of wastewater services, as distinct from freshwater services, no place exists for these in a free market society, except in respect of internal and external re-use. *Laissez-faire* is inherently incapable of dealing with pollution, because pollution remission is not a privately appropriated good. The social service society – the emphasis here must be that this is a theoretical construct – *does* regulate water pollution, as part of its basic commitment to public good provision.

In real societies, where only intermediate institutional forms exist, the scale of pollution is a function of the rate of economic activity, and of domestic, agricultural and industrial technologies – both as perpetrators and as remitters of pollution. The scale of pollution also depends on the environmental consciousness and organization of individuals, of households, of the institutions of civil society such as NGOs, churches and political parties, of farmers, of the managers of companies and of government. Pollution is limited by wastewater treatment, the costs of which are borne by the domestic and industrial sectors through the price of water, or fixed charges, or discharge payments. Pollution limitation is also delivered by regulations that ban specific agricultural and industrial practices or by a requirement for industry to treat its effluent prior to disposal. But pollution remission can be high cost, in financial terms, and for that reason it meets widespread resistance.

9.5 Hydroeconomics and institutional policy

In Chapter 1, it was stated that the paradigm from which the analysis is developed is that of evolutionary political economy. Enquiry in evolutionary and institutional economics is value-driven, exhibits a concern for the improvement of the human condition, seeks to relate analysis to policy, recognizes the centrality of participatory democratic processes to the identification and evaluation of real needs, aims to achieve social reform through collective action constructively to recast institutional forms, but abjures policy universals.

It follows, then, that this concluding chapter should embody the principal policy recommendations of Chapters 2–8. I hope that these proposals will be of use to a broad range of institutions such as NGOs, water companies in the public or private sector, regulators, local, regional and national governments, and international bodies. But all of these proposals should be critically reviewed in the context of each specific catchment, region or country.

The substantive mission of water resource policy should be to supply water of sufficient quantity and appropriate quality to users in households, agriculture, industry and other sectors; to ensure the use of fresh water is affordable to low-income households; to ensure the husbandry of water in its supply and use; to purify water from domestic, agricultural and industrial effluents; to prevent the abuse of monopoly power in the supply of fresh water and the collection of wastewater; to protect rural and urban communities against floods and to drain the land of stormwater; to protect water's hydrocyclical capacity to renew its ground- and surface water flows; to conserve natural species and habitats in all their fresh- and coastal-water environments; to reduce and eliminate water-driven international conflict; and to ensure that, when government expenditure takes place for these purposes, it is spent wisely.

It may be useful to set out strategic policy options for a catchment, a region, a country or a group of countries using Table 9.1, which is developed from the water balance statement of Table 2.1

Table 9.1 would show, for each form of supply, a specific option for a planning period of, say, five or ten or fifteen years. Further versions of Table 9.1 could show

Table 9.1 Strategic quantitative options for water resource planning.

Forms of supply	Change in quantity	Users	Change in quantity
Ground water sources	a	Households	t
Surface water sources (including tidal waters)	b	Agriculture, forestry	u
Desalination of salt or brackish waters	c	Mining	v
Import of water from another catchment	d	Manufacturing	w
Internal re-use of waste water	f	Public services	x
External re-use of waste water	g	Commercial sectors	y
Leakages	h		
Less: export of water to another catchment	j	Other users	z
Total change in net supply a+b+c+d+f+g+h–j		**Total change in use** t+u+v+w+x+y+z	

Note: Total net supply change equals total use change.

alternative options. The values from a to j indicate the quantitative change in supply over the planning period – say ten years. Values t to z indicate the quantitative change in consumption by each user group over the same ten years. Total change in net supply is equal to total change in consumption. Where a form of supply or use increases, it is written as a positive number. If supply or use diminishes, it has a negative value. Note that Table 9.1 assumes leakage losses will fall, so this is written as $+h$ in the total net supply row, since it is equivalent to a supply expansion. Calculations should be made, for both the base period and the end of the ten year planning period, of the catchment's self-sufficiency and its stress indicator (see §2.8). The strategy would also show *qualitative* change. This can be done by taking each use-class and setting out how the quality of the water it receives might change and, secondly, how the quality of the wastewater that it releases might change.

With respect to changes in supply, it is important that policy-makers consider the full range of options open to them. These can include:

- the erection of new infrastructures to abstract fresh water from sources below the ground or from those on the land's surface
- abstraction from estuaries and the sea
- the expansion of existing abstraction plant, to raise installed capacity
- infrastructural works and institutional restructuring to raise usable capacity closer to theoretical capacity
- the reassignment of authorizations to abstract between licensees, thereby increasing the rate of abstraction within a single catchment area
- the trading of abstraction rights
- more sophisticated control in the management of existing reservoirs
- more effective conjunctive use of rivers, underground sources and reservoirs
- the expansion of the storage network

- investment and changed management practices to reduce water losses from existing supply networks
- inter-regional transfers, where a shortage of demand with respect to supply exists in one region and a surplus in another
- the extension of the internal re-use of water
- the extension of the external re-use of water
- the provision of infrastructures to extend recycling to raise the measure of the "corrected supply" available for abstraction (see Eq. 2.5).

Changes in the quantity of water supplied to users, and improvements in both the quality of the water they receive and the quality of water recycled or released into the sea, all require substantial financial resources. It is vital that, in the regulation of water monopolies, sufficient funds are left within the sector to make possible the investments that public policy requires.

At this point, one turns the spotlight from the supply side to questions of use. During the past two or three decades, widespread and intense competition between users for scarce water has led to strong interest in a battery of policies to reduce the quantity of water consumed at any given level of economic development. These innovations have become known as demand management and they comprise internal and external re-use, consumption technology, land-use planning, educational initiatives, sectoral adjustment in dependence on imported food, and water pricing (see §4.5).

Water pricing requires investment in volumetric metering and should be considered as complementary with other demand-management measures, not as a substitute for them. The absence of pricing leads to the waste of the resource, creating unnecessary economic and environmental costs. Tariffs for fresh- and wastewater services should have four objectives. They must be affordable to the households purchasing these services; they should cover fully the prime and overhead costs of service production, thereby eliminating the dependence of the water enterprise on local or national government subsidy; they should underpin the quest for sustainability by encouraging water conservation; and they should stimulate the protection of the environment from pollution. The first of these objectives may conflict with the other three; this inconsistency can be resolved by a combination of low pricing for households' basic water needs and government subsidy targeted to families in special needs.

In evaluating the economic outcome of projects, social cost–benefit analysis and social cost-effectiveness analysis, incorporating shadow prices, are valuable techniques for making public choices. Where projects are independent, their net present value should be used; where they are interdependent for technical reasons or because of a budget constraint, the net benefit–investment ratio is recommended. The social time-rate of discount used by national government should not be confused either with market rates of interest or the opportunity cost of capital, and the consistency of a low social time-rate of discount with a sustainability agenda should be understood.

The incorporation of environmental impacts into project analysis should be by means of dual analysis. Dual analysis refers to the parallel appraisal of a project by social cost–benefit analysis and environmental impact analysis. The final choice between project options uses both sets of results, economic and environmental, including their distributional outcomes, but recognizes that they are not commensurable. The variant costing method is also useful within dual analysis. The environmental impact statement is used to specify project variations by which negative environmental effects can be mitigated or positive effects reinforced: each variation changes the measured costs of a project while changing the nature of its environmental effects, so each of the variations constitutes a distinct project option.

In respect of both project analysis and financial accounting appraisal, the leverage for bias must be contested. In the tendering process for feasibility studies, clients and shadow clients of consulting firms should select only on the basis of the quality and cost of competing bids and be open-minded about the methods most appropriate for the research. Client and shadow client should not seek to lever the work, once it is under way, towards outcomes they favour. Where aid projects are initiated through bilateral client–contractor negotiations, and where they promise substantial exports for the aid-giving country, special attention must be given to protecting the evaluation process from commercial and governmental pressures. Accountants should practise as well as preach the ethical conventions of consistency, objectivity and relevance. Integrity consulting should be supported by the professional associations in water and environmental management, such as the UK's Chartered Institution of Water and Environmental Management, and these associations should also adopt a set of ethical principles.

With respect to the legal form of water utilities, there should be no policy universals. Government, citizens and the institutions of civil society should seek an intermediate set of relations, between the social service and the free market ideal types, which best reflects the country's specific needs.

Some form of countervailing power is required to represent the common good in the regulation of the water industry, whether the monopolies be in the private or the public sector. Environmental regulation should be institutionally separate from economic regulation, a necessary condition for creative tension between these processes. When a utility delivers both a core fresh- and wastewater service and pursues enterprise activities or non-water business, the core service should be legally and financially separate and, in the case of public limited companies, have a separate listing on the stock exchange (Byatt 1996).

The economic regulation of water sector monopolies by government or a government agency could comprise the following activities:

Abstraction
- regulating the charges payable by the monopolies to the state (or another institution) for water abstraction
Distribution
- regulating the percentage losses of water through leakage
Freshwater use
- regulating the charges payable by users to the monopolies for freshwater supply

- regulating the use of metering in estimating users' water consumption
- regulating the level of services in freshwater supply provided by the monopolies to users, with respect to pressure, security of supply, and so on

Land drainage

- regulating the charges payable by local populations for land drainage, and the level of services in land drainage, where such charges and services are provided by a monopoly

Wastewater services

- regulating the charges payable by water users to the monopolies for wastewater collection, treatment and disposal
- regulating the charges payable to the state by industry – in the broadest sense – for the right to discharge effluents into surface and groundwater sources

Water sector monopolies

- regulating the dividends payable by the monopolies to their shareholders
- regulating the rate of profit on capital earned by the monopolies
- regulating the level of payments and pensions by the monopolies to their directors and managers.

Government should set abstraction charges and discharge fees at least at a level to recoup capital and current expenditures on the environmental and economic regulatory processes. Moreover, abstraction charges borne by water monopolies or by other abstractors should incorporate incentive charging, varying the charge according to the volume of water taken, the location of the abstraction, the time of the abstraction, and the amount of water likely to be returned directly to the water course. Similar principles should be adopted for discharge payments.

In the design of economic regulation, government should consider the following: the US practice of a robust and independent regulatory commission, regular price and service reviews, with space for additional review where appropriate, openness on the part of the regulator with regard to the basis of its decisions, and parliamentary scrutiny of the regulatory process.

In the process of settling the intermediate institutional character of its water industry, options to pursue some form of public sector provision of the core service should not be deemed automatically to impose a financial burden on government. No such burden exists where public provision is through a public trading enterprise, where the general government financial deficit is the key fiscal target of government, where the public trading enterprise is well run, and where its profit and loss account is in surplus.

On a global scale, irrigated agriculture is a major user of abstracted water. Much of this consumption is wasteful and can be modified by infrastructural improvements to reduce leakages, fiscal subsidy to fallow poor quality agricultural land, reduced subsidies for water-inefficient crops, trading in abstraction rights, and full-cost water pricing. But where such policies raise a country's dependence on food imports, such as grains, careful consideration needs to be given to the long-term trend in the price of imported food.

In the late 1990s, water resource policy is sometimes seen as ripe for a shift from supply-side dominance to an era of demand management. I believe this is a mistaken approach. My point of view is that water resource strategies must first be set within the much broader and ambitious goal of a sustainable society. Within the world of water, such strategies should embody the six fields of action described in Chapter 7 as constituting the sustainability principle for water resource management. These six fields are:

- protection of the hydrocycle
- purification of water
- conservation of species and habitats
- husbandry of water in its supply and use
- the efficient supply of fresh water to humankind
- the drainage of land and protection against floods.

They require clear-sighted action, in respect of both supply and use. Water resource policy must walk on two legs.

Finally, I hope the reader agrees that the supply of sufficient fresh water and the provision of wastewater services, as well as the protection of the natural supplies of unpolluted water to the animal and plant life of the Earth, are one of the great vocations for the third millennium. Understanding water resources requires the skills of hydrologists, engineers, economists, environmentalists, political scientists, lawyers and many others. This introduction to the economics of water resources has sought to make evolutionary political economy a servant to them all.

GLOSSARY

abstraction   The act of abstracting or taking away.

accumulation of capital   Growth of real and financial assets, particularly of a firm.

activated sludge treatment   An aerobic biological process for conversion of soluble organic matter to solid biomass subsequently removable by gravity or filtration.

ambient   Surrounding.

ambient source   Sources distributed widely in environmental media.

amortization   Procedure of a firm setting aside funds to replace capital equipment when it becomes obsolete.

aqueduct   An artificial channel for conveying water, especially noticeable when in the form of a bridge supported by tall columns across a valley.

aquifer   An underground rock formation which is able both to store and yield significant quantities of groundwater.

assets   Real and financial wealth.

ATCmin   Minimum average total cost.

average overhead cost   Total overhead cost divided by the number of units of output.

average prime cost   Total prime cost divided by the number of units of output.

average total cost   Total cost divided by the number of units of output.

backloading   An effect whereby costs are incurred later in their time-stream.

balance sheet   A statement of a firm's assets and liabilities.

biochemical oxygen demand   A measurement of the quantity of oxygen used in the biochemical oxidation of carbonaceous and nitrogenous organic compounds in a specified time, at a specified temperature and under specified conditions; the standard measurement is made for five days at 20°C and is termed BOD5; an indicator of the presence of organic material in water.

borehole   A deep, narrow hole, especially one drilled into the Earth to monitor or exploit water, oil, gas, etc.

brackish   Slightly salty.

capital account spending   See capital expenditure.

capital expenditure   Expenditure where the resource purchased has an expected life of more than 12 months.

cashflow statement   A financial statement setting out for a given time-period all of a firm's cash inflows and outflows.

catchment (area)   The land area from which all rainfall eventually flows into a specified river.

collection   In this book, the networked retrieval of storm and foul water.

conjunctive use   The planned connective deployment of discrete water resources.

constant prices   A fixed set of prices applied over more than one time-period.

consumer's surplus   The notion that, with a downward-sloping demand curve, a single market price understates the value to consumers of intramarginal consumption.

consumption   See **use**.

consumptive use   Use of water such that little is recycled.

contradiction   The development of forces within a social formation which threaten the long-term expansion and stability of that society.

cryptosporidium   A microscopic parasite that can cause diseases in humans.

cubic function   In this context, a curve that first falls steeply, then gently, then steeply again.

culvert   An underground channel routing a watercourse under a road, development, etc.

current account spending   See current expenditure.

current expenditure   Expenditure on services, labour, and goods with a life of 12 months or less.

dam   An engineered barrier constructed to hold back water and raise its level, forming a reservoir or preventing flooding.

debenture   A specialized loan with a trustee to look after the holders' interests.

demand   See **effective demand**, and **use**.

demand management   Policies to reduce the quantity of water that users choose to consume.

depreciation   An accounting device to write down the value of a firm's capital goods.

desalination   A process for removing inorganic salts from brackish, saline, or sea water.

disposal   In this book, the process of getting rid of wastewater and sludge.

dissolved oxygen   Just as a solid – such as salt – is soluble in water, so is the gas oxygen; the quantity of oxygen dissolved in water is a measure of water quality, indicating its capacity to support life.

distribution   In this book, the delivery of fresh water to users.

diurnal   Daily (24 hours).

dividends   Moneys paid in respect of share ownership.

dual supply   The supply of water of different qualities to different categories of user.

economies of scale   Lower average total cost at higher output levels.

effective rainfall   Total rainfall in an area minus that lost through **evapotranspiration**.

environmental costs   Negative ecological impacts.

equity   The excess of a firm's assets over its liabilities.

estuary   A wide, tidal mouth of a river.

eutrophic (of a water body)   Rich in nutrients and therefore supporting a dense algal or plant population, which ultimately kills animal life by depriving it of oxygen.

evaporation   The loss of water moisture as a vapour, excluding **transpiration**.

evapotranspiration   The combined processes of **evaporation** and **transpiration**.

evolutionary political economy   An approach to economics notable for its evolutionary character, its institutionalism, its epistemology, its rootedness in ethical values, its open-mindedness and its concern for the interdependence of economy and environment.

ex ante   An adjective meaning prior in time.

ex post   An adjective meaning after in time.

extended social cost–benefit analysis   Social cost–benefit analysis extended by means of the monetization of environmental impacts that have no clearly measurable gross domestic product effect.

external re-use The re-use of water outside the institution in which it was first used.

externality The environmental and economic impact of a project on individuals or institutions that neither work in the project nor use its outputs.

fiscal Concerning the getting and spending of money by government.

fossil aquifer A deep **aquifer** formed in prehistoric times and no longer receiving a significant recharge.

foul sewer A sewer for the collection or disposal of foul water.

foul water Water polluted by prior use.

fresh water Water that is not salty.

frontloading An effect whereby costs are incurred earlier in their time-stream.

gravity flow Water flow caused by the force of gravity.

gross domestic product The total value of the goods and services produced by the residents of a country.

gross margin The difference between unit price and average prime cost at any specific output level.

groundwater That part of the natural water cycle present within underground strata and aquifers.

groundwater overdraft The excess of groundwater **abstraction** over long-term **recharge**.

headworks costs The costs of **abstraction, storage** and **treatment**.

hydraulic (of water, oil, etc.) Conveyed through pipes or channels usually by pressure; (of a mechanism etc.) operated by liquid moving in this manner.

hydraulics The science of the conveyance of liquids through pipes, etc., especially as motive power.

hydroeconomics The economics of water resources.

hydrological cycle The natural cycle of precipitation, interception, ground- and surface water flows, **evapotranspiration** and condensation.

hydrosocial cycle The social processes of freshwater **abstraction, storage**, treatment, **distribution, use**; and **wastewater collection**, treatment and **disposal**; as well as the cyclical activities of internal re-use, external re-use and recycling.

hyperbolic function A function in which a variable's value approaches but never reaches a limiting value.

ideal type An idealized model of human interaction.

income elasticity of demand The proportionate difference in the quantity purchased of a commodity between households in two income groups divided by the proportionate difference in their incomes, i.e.:

$$\frac{Q_A - Q_B}{Q_A} \Bigg/ \frac{Y_A - Y_B}{Y_A}$$

indivisibility The idea that, to secure even modest levels of output, major works are necessary.

infiltration   The process of water entry into the soil through the soil surface.

inquination   In this book, the four processes of destruction, depletion, degradation and disablement of the environment.

installed capacity   The output that it is theoretically possible to deliver with the existing technical infrastructures.

internal rate of return   The discount rate at which the discounted value of incremental real output precisely equals the discounted value of incremental real costs.

internal re-use   The re-use of water within an institution in which it has already been used.

irrigation   The **abstraction**, **distribution** and **use** of water in agriculture.

land drainage   The process of drawing water off the land.

landfill   The use of land sites for the disposal of solid waste.

latrine   A communal lavatory, especially in a camp, barracks, etc.

liabilities   The debts of a firm.

long term   A time-period sufficient to install additional capital equipment, either in new projects or for the expansion of existing infrastructure and plant.

lumpiness   In respect of infrastructure, a characteristic such that capital equipment installed today raises the cost of similar equipment installed tomorrow.

mains (or main)   Principal channel for water, sewage, etc.

marginal cost   Of the nth unit of output is defined as the difference in total cost between producing n and $(n-1)$ units.

marginal revenue   For the nth unit of output is defined as the difference in total sales income derived from selling n rather than $(n-1)$ units.

modal income   The most common income received within a population.

monopsony   The exclusive purchaser in a market.

net benefit–investment ratio   The net present value for a project's working life divided by the net present value of its gestation period.

net present value   The present value of the net benefit stream discounted at the social time-rate of discount.

net profit   Total sales value less total costs.

network costs   The costs of freshwater **distribution**, stormwater and foul water **collection**, and **wastewater disposal**.

non-consumptive use   Use of water such that the bulk is recycled.

opportunity cost   The output value forgone in GDP terms if a productive resource is deployed for a defined project or use, rather than its existing use.

opportunity cost of capital   The return on financial capital forgone if capital is deployed for a defined project rather than in its alternative use.

outfall   The mouth of a river, drain, etc., where it empties into the sea, etc.

out-turn prices   The ruling set of prices for each time-period.

overdraft   An agreed limit by which a firm's bank current/chequing account can be in debit.

overhead costs   A firm's costs other than prime costs, often assumed to be fixed whatever the level of production.

pathogen   An agent causing disease.

percentile   If the elements of a set are ranked from top to bottom, then grouped into one-hundredth parts, the *n*th percentile is the *n*th group from the bottom.

pH   A measure of acidity or alkalinity.

pre-treatment   The treatment of **wastewater** by a user, prior to its disposal.

primary treatment (of wastewater)   First-stage treatment by means of screening, grit removal and settlement of organic solids.

prime cost   Resources used up by a firm in the daily production of goods and services, consisting of the salaries and wages of the workforce, the costs of power, materials, spare parts, and other consumables such as bought-in specialist inputs and services.

principal   The original debt itself, as distinct from the interest on the debt.

profit and loss account   A financial statement containing entries that show for an accounting year whether a firm made a profit or a loss from its activities during that year.

public good   Any good or service where the benefit derived from it by one consumer does not reduce the benefit derived by consumers in general.

public limited company   A profit-seeking private company, the shares in which are available for sale to the general public and in which the owners' liability, should the company be bankrupted, is limited to the loss of their share capital.

quadratic function   In this context, a curve which first falls, then rises.

recharge   Water for a renewed charge (usually of **aquifers**) by natural or man-made means.

recycling   The release of **wastewater**, with or without prior treatment, into surface water and **groundwater**, where it supplements the natural flows down stream from its point of disposal.

reservoir   A large natural or artificial lake used as a stored source of water supply.

retained profits   Net profit, less taxation and dividend payments to shareholders.

retrofit   To fit new equipment into a building or machine after its original construction.

riparian   Of or on a river bank.

runoff   Rainfall that is carried off an area by streams and rivers or from urban surfaces through drains, etc.

safe yield (of installed capacity)   The flow of water available, for example, during a 1 in 50 year drought.

secondary biological treatment (of wastewater)   Settlement of fine organic particles and subsequent anaerobic digestion to form methane gas, and a stable, solid and oxidative treatment of the separated liquor, either by activated sludge filtration or forced air oxidation processes.

septic tank   A tank in which the organic matter in sewage is slowly denatured through anaerobic bacterial activity; the liquid continuously overflows into the ground, but residual solids remain to be periodically desludged.

sewage   Waste matter, especially excremental.

sewage farm/works   A place where sewage is treated, including the production of sewage sludge.

sewer   A conduit, usually underground, for carrying off drainage water and **sewage**.

sewerage   A networked system of **sewers**.

shadow prices   Non-market prices, used to reflect the opportunity cost of a resource.

short term   A time-period in which increased daily output is possible only through operational changes.

silage   Green fodder that has been stored in a silo.

sludge   Residue left after treating storm water or foul water.

slurry (in farming)   A fluid form of manure.

soakaway   An arrangement for disposing of **wastewater** by letting it percolate through the soil.

social time-rate of discount   A variable employed to reduce the value of project benefits and costs in proportion to the time-delay before the benefit or the cost is registered, reflecting a society's uncertainties about the future.

standpipe   A vertical pipe extending up from a piped water supply, especially one connecting a temporary tap to the underground mains.

storage   The use of **reservoirs**, etc., to retain water for later use.

storm sewer   Sewer drains used for the collection of stormwater.

stormwater   Rainfall runoff, especially during intense precipitation.

supply   Generally, the production and distribution of goods and services; with reference to the supply of **fresh water**, it is the **abstraction**, **storage**, **treatment** and **distribution** of fresh water, including the desalination of salt water.

surface water   The water found in rivers, inland seas, lakes and pools.

suspended solids   Non-soluble materials in water, removable by settlement and filtration.

sustainability   See §7.3.

switchgear   Machinery for converting the voltage of a supply of electricity.

tailwater   Water exiting from irrigated land.

tariff   The price of a good or service, or a set of such prices.

tertiary treatment (of wastewater)   Additional treatment of the oxidized, settled liquor through one or more "polishing" processes involving filtration and/or oxidative retention prior to final disposal.

theory of differential rent   Ricardo's theory that the rent of land varies with its natural fertility.

total costs   The sum of **prime costs** and **overhead costs**.

tradable abstraction rights   Legal rights to abstract water; they can be bought and sold.

trade credit   The period of grace conceded by materials and equipment suppliers between the date their goods are delivered and the date by which payment is required for them.

transpiration   The release of water vapour from plants and trees.

treatment   In this book, the mechanical, chemical and biological processes that raise the quality of fresh water and waste water.

turnover   The value of sales per unit period of time.

unaccounted-for-water   Discrepancy between water flows leaving the works and the total sum of all water received by consumers; although mainly leakages, it includes significant metering errors and unknown/illegal diversions.

usable capacity   The maximum output practically possible.

use   In this book, used synonymously with consumption, to mean the application of water to a purpose by households, agriculture and industry.

virtual water   The water used to grow exported food.

wadi   A rocky, ephemeral watercourse, dry except in the rainy season.

wastewater   **Stormwater** and **foul water**, more specifically foul water.

water balance statement   A categorization of the forms of supply and use of abstracted water, for example in a catchment, and the quantification of their volumes per unit period of time.

water losses   The leakage and evaporation of water within the **hydrosocial cycle**; or, specifically, leakages in the supply system.

water productivity   A sector's production of goods and services (measured in value-added terms) divided by water used (measured in m^3).

water table   A level below which the ground is saturated with water.

water treatment plant   An infrastructure for the cleaning of **foul water** and **stormwater**.

weir   A small dam built across a river to raise the level of water up stream and/or regulate its flow

REFERENCES

Adams, W. & J. Brock 1994. Corporate performance. See Hodgson et al. (1994: vol. I, 100–103).

Åkerman, J. 1960. *Theory of industrialism: causal analysis and economic plans*. Lund: Lund University Press.

Allan, J. A. 1995a. The political economy of water: reasons for optimism but long term caution. See Allen & Court (1995: 33–58).

— (ed.) 1995b. *Water and the Middle East peace process: negotiating water in the Jordan basin*. London: I. B. Tauris.

Allan, J. A. & J. H. O. Court (eds) 1995. *Water in the Jordan catchment countries: a critical evaluation of the role of water and environment in evolving relations in the region*. London: School of Oriental and African Studies.

Allan, J. A. & M. Karshenas 1995. Managing environmental capital: the case of water in Israel, Jordan, the West Bank and Gaza, 1947–1995. See Allan (1995b: ch. 6).

Allen, R. E. 1990. *The concise Oxford dictionary of current English*. Oxford: Oxford University Press.

Amin, A. 1994. Corporate concentration and interdependence in Europe. See Hodgson et al. (1994: vol. I, 85–91).

Andreoni, J. 1990. Impure altruism and donations to public goods: a theory of warm-glow giving. *Economic Journal* **100**, 464–77.

Anonymous 1995. Hydroelectric system Gabcíkovo–Nagymaros. *Business Europa* (February–March), 24–6.

Arlosoroff, S. 1995. Managing scarce water – recent Israeli experience. See Allan & Court (1995: 21–8).

Bateman, I. 1995. Environmental and economic appraisal. See O'Riordan (1995: 45–65).

Bateman, I., I. Langford, A. Graham 1995. *A survey of non-users' willingness to pay to prevent saline flooding in the Norfolk Broads*. Working Paper GEC 95–11, CSERGE, University of East Anglia.

Bausor, R. 1994. Time. See Hodgson et al. (1994: vol. II, 326–30).

Binder, J. 1993a. Gabcíkovo after nine months. *Slovakia* **1**(1), 38.

Binder, J. 1993b. Confirmed hopes. *Slovakia* **I**(5), 26–7.

Björnlund, H. 1995. Transferable water rights – Australian application. Unpublished paper, Faculty of Business and Management, University of South Australia.

Björnlund, H. & J. McKay 1995. Can water trading achieve environmental goals? *Water* (November/December), 31–4.

Byatt, I. 1996. The case for an amicable separation. *Financial Times* (9 January).

Carson, R. 1962. *Silent spring*. Boston: Houghton Mifflin.

CEC 1985. *Council directive on the conservation of wild birds 79/409*. Brussels: CEC.

— 1994. *Latvia municipal services development project: Daugavpils Water and Sewerage Enterprise – Udensvads*. Brussels: CEC.

DOE (Department of the Environment) 1991. *Policy appraisal and the environment: a guide for government departments*. London: HMSO.

Department of Water Resources 1994. *California water plan update* [Final Bulletin, 160–93]. Sacramento: Californian Department of Water Resources.

Desvousges, W. H., V. K. Smith, A. Fisher 1987. Option price estimates for water quality improvements: a contingent valuation study of the Monogahela River. *Journal of Environmental Economics and Management* **14**, 248–67.

Diamond, P. A. & J. A. Hausman 1994. Contingent valuation: is some number better than no number? *Journal of Economic Perspectives* **8**(4), 45–64.

Dyson, J. R. 1994. *Accounting for non-accounting students*, 3rd edn. London: Pitman.

Dyson, T. 1994. Population growth and food production: recent global and regional trends. *Population and Development* **20**(2), 397–411.

EAEPE 1991a. Theoretical perspectives. *EAEPE Newsletter* (5), 20.

— 1991b. Rigour and pluralism in economics. *EAEPE Newsletter* (6), 1.

ECOTEC 1993. *A cost–benefit analysis of reduced acid deposition: UK natural and semi-natural ecosystems*. Birmingham: ECOTEC.

EIU 1994. *Baltic republics: Estonia, Latvia, Lithuania: country profile 1994–95*. London: Economist Intelligence Unit.

— 1995a. *Baltic republics: Estonia, Latvia, Lithuania: country report 1st quarter 1995* London: Economist Intelligence Unit.

— 1995b. *Peru: country profile 1994–95*. London: Economist Intelligence Unit.

EPDRB Task Force 1995. *Strategic action plan for the Danube river basin 1995–2005*. Vienna: Environmental Programme for the Danube River Basin.

Evans, H. R. 1993. The structure and management of the British water industry 1945–91. *IWEM Yearbook 1993*. London: CIWEM.

Foster, J. 1994. Biology and economics. See Hodgson et al. (1994: vol. I, 186–93).

Fowler, S. 1995. *Water wise: the RSPB's proposals for using water wisely*. Sandy: RSPB.

Gindlova, D., I. Kusy, S. Merrett, D. Petrikova 1995. *On law and practice: environmental legislation in Slovakia*. Bratislava: Slovak Technical University.

Gittinger, J. P. 1982. *Economic analysis of agricultural projects*. Baltimore: Johns Hopkins University Press.

Glasson, J. G., R. Therivel, A. Chadwick 1994. *An introduction to environmental impact assessment*. London: UCL Press.

Gleick, P. H., P. Loh, S. V. Gomez, J. Morrison 1995. *California water 2020: a sustainable vision*. Oakland: Pacific Institute for Studies in Development, Environment, and Security.

Goldsmith, E., R. Allen, M. Allaby, J. Davŏll, S. Lawrence 1972. Blueprint for survival *The Ecologist* **2**(1), 1–40.

Gordon, M. 1993. Water pressures – a time for planning. *Town and Country Planning* **62**(9), 236–40.

Gough, I. 1994. Concept of need. See Hodgson et al. (1994: vol. II, 118 –26).

Graham, G. 1995. Problem loans trail World Bank. *Financial Times* (25 September).

Green, C., S. Tunstall, E. Penning-Rowsell, A. Coker 1990. *The benefits of coast protection: results from testing the contingent valuation method for valuing beach recreation.* London: Middlesex Polytechnic.

Gustafsson, M. 1993. *From biocides to sustainability: Swedish environmentalism 1962– 1992.* Örebro: University of Örebro.

Hearne, R. R. & K. W. Easter 1995. *Water allocation and water markets: an analysis of gains-from-trade in Chile.* Washington DC: World Bank.

Heidebrecht, R. & N. Hewitt 1994. Environmental management of water. In *Guide to environmental management for local authorities in Central and Eastern Europe*, K. Otto Zimmermann, ch. 12. Freiburg: International Council for Local Environmental Initiatives.

Hills, J. S. 1995. *Cutting water and effluent costs*, 2nd edn. London: Institution of Chemical Engineers.

Hirschleifer, J., J. C. DeHaven, J. N. Milliman 1960. *Water supply: economics, technology and policy.* Chicago: University of Chicago Press.

HoCCPA (House of Commons Committee of Public Accounts) 1994. *Pergau hydroelectric project.* London: HMSO.

HoCFAC (House of Commons Foreign Affairs Committee) 1994. *Public expenditure: the Pergau hydroelectric project, Malaysia, the aid and trade provision and related matters.* [2 volumes]. London: HMSO.

Hodgson, G. 1994a. Theories of economic evolution. See Hodgson et al. (1994: vol. I, 218– 24).

— 1994b. Cultural and institutional influences on cognition. See Hodgson et al. (1994: vol. I, 58–63).

— 1994c. Determinism and free will. See Hodgson et al. (1994: vol. I, 134–8).

— 1994d. Habits. See Hodgson et al. (1994: vol. I, 302–305).

Hodgson, G., W. Samuels, M. Tool (eds) 1994. *The Elgar companion to institutional and evolutionary economics* [2 volumes]. Cheltenham: Edward Elgar.

Hufschmidt, M., L. Fallon, J. Dixon, Z. Zhu 1987. *Water management policy options for the Beijing-Tianjin Region of China.* Hawaii: East–west Center.

Hughes, G. 1991. Cost–benefit analysis: housing and squatter upgrading in East Africa. In *Housing the poor in the developing world: methods of analysis, case studies and policy*, A. G. Tipple & K. Willis (eds). London: Routledge.

Hutton, W. 1995. Why Treasury needs a lesson in counting. *The Guardian* (1 May).

Jackson, H. 1995. Revamping water regulation. *New Ground* (Winter), 7–8.

Kapp, W. 1983. *Social costs, economic development and environmental disruption.* Lanham: University Press of America.

Kay, J. 1994. Clever trick but the cracks remain. *Financial Times* (16 August).

Kay, N. 1994. Theory of the firm (II). See Hodgson et al. (1994: vol. I, 237–41).

Kinnersley, D. 1994. *Coming clean: the politics of water and the environment.* London: Penguin.

Khalil, E. 1994. Entropy and economics. See Hodgson et al. (1994: vol. I, 23–9).

Krejci, P. 1994. Slovnaft: environmental care. *Slovakia* II(5), 18–19.

Kundera, M. 1984. *The unbearable lightness of being.* London: Faber & Faber.

Lee, F. 1994a. Administered prices. See Hodgson et al. (1994: vol. I, 4–9).
— 1994b. Full cost pricing. See Hodgson et al. (1994: vol. I, 262–6).
Le Heron, R. 1993. *Globalized agriculture*. Oxford: Pergamon Press.
Lipton, M. 1992. The spectre at the fast. *Financial Times* (24 June).
Little, I. M. G. & J. A. Mirrlees 1991. Project appraisal and planning twenty years on. In *Proceedings of the World Bank Annual Conference on Development Economics 1990*, World Bank. Washington DC: The World Bank.
Lundqvist, J., U. Lohm, M. Falkenmark 1985. *Strategies for river basin management: environmental integration of land and water in a river basin*. Dordrecht: Reidel.

Malpezzi, S. 1990. Urban housing and financial markets: some international comparisons. *Urban Studies* **27**(6), 971–1022.
Manson, J. 1994. At the limits: the success and failure of water privatization in Britain. *The Urban Age* **2**(4), 11–12.
Månsson, T. 1992. *Ecocycles: the basis of sustainable urban development*. Stockholm: Statens Offentliga Utredningar.
Marshall, A. 1962. *Principles of economics: an introductory volume*. London: Macmillan.
McCann, W. & B. Appleton 1993. *European water: meeting the supply challenges*. Camborne: Financial Times Management Reports.
McDonald, A. & D. Kay 1988. *Water resources: issues and strategies*. Harlow: Longman.
Merrett, S. 1971. Snares in the labour productivity measure of efficiency: some examples from Indian nitrogen fertiliser manufacture. *The Journal of Industrial Economics* **XX**(1), 71–84.
— 1979. *State housing in Britain*. London: Routledge & Kegan Paul.
— 1989. *Rational financial choice between alternative housing time-paths: a net future value approach*. Housing Finance Discussion Paper 1, Joseph Rowntree Trust, York.
— 1994. *The evaluation of housing projects in the developing countries*. Working Paper 9, Planning and Development Research Centre, University College London.
— 1995. Planning in the age of sustainability. *Scandinavian Housing and Planning Research* **12**, 5–16.
— 1997. *Joint venture projects in water infrastructure: finance, risk and contract*. London (mimeo).
— (with F. Gray) 1982. *Owner occupation in Britain*. London: Routledge & Kegan Paul.
Merrett, W. J. 1992. *Hydrological source area liming to ameliorate surface water acidification*. Cardiff: Welsh Region National Rivers Authority.
Mjøset, L. 1994. Johan Åkerman. See Hodgson et al. (1994: vol. I, 9–11).
Munasinghe, M. 1990. *Managing water resources to avoid environmental degradation: policy analysis and application*. Washington DC: World Bank.

Naess, A. 1989. *Ecology, community and lifestyle: outline of an ecosophy*. Cambridge: Cambridge University Press.
NAO (National Audit Office) 1993. *Pergau hydroelectric project*. London: HMSO.
Nelson, R. 1994. Theory of the firm (II). See Hodgson et al. (1994: vol. I, 241–6).
Neutze, M. 1994. The costs of urban physical infrastructure services. Working Paper 42, Urban Research Program, Australian National University.
— 1997. *Funding urban services*. Sydney: Allen & Unwin.
NRA 1995. *1995–96 annual abstraction charges*. Bristol: National Rivers Authority.

NRA Thames Region 1994. *Future water resources in the Thames region: a strategy for sustainable management*, Reading: National Rivers Authority Thames Region.

— 1995. *Fact file 7/7: water resources*. Reading: National Rivers Authority Thames Region.

— undated. *Fact file 2/7: flood defence*. Reading: NRA Thames Region.

Ofwat 1993. *Paying for quality: the political perspective*. Birmingham: Ofwat.

— 1994a. *Water supply, sewage disposal and the water environment: a guide to the regulatory system*. Birmingham: Ofwat.

— 1994b. *The urban wastewater treatment directive*. Birmingham: Ofwat. Information Note No.24.

— 1994c. *1993–94 report on the cost of water delivered and sewage collected*. Birmingham: Ofwat.

— 1995. *1994–95 report on the cost of water delivered and sewage collected*. Birmingham: Ofwat.

O'Riordan, T. (ed.) 1995. *Environmental science for environmental management*. Harlow: Longman.

Pearce, D., A. Markandya, E. Barbier 1989. *Blueprint for a green economy*. London: Earthscan.

Penning-Rowsell, E., C. Green, P. Thompson, A. Cokes, S. Tunstall, C. Richards, D. Barker 1992. *The economics of coastal management: a manual of benefit assessment techniques*. London: John Wiley.

Pigram, J. J. 1986. *Issues in the management of Australia's water resources*. Melbourne: Longman.

Pigram, J. J., R. J. Delforce, M. L. Coelli, V. Norris, G. Antony, R. L. Anderson, W. F. Musgrave 1992. *Transferable water entitlements in Australia*. Armidale: Centre for Water Policy Research.

Pinkham, R. 1994. *Improving water quality with more efficient irrigation*. Snowmass: Rocky Mountain Institute.

Plender, J. 1996. FT guide to privatisation. London: *Financial Times* (8 January).

Powers, T. & C. Valencia 1978. *Modelo de simulación de obras públicas (SIMOP): manual de usuario*. Washington DC: Inter-American Development Bank.

PRONAP 1994. *Concurso internacional de meritos para la contratacion de firmas consultoras: proyectos de mejoramiento institucional y operativo: terminos de referencia*. Lima: Ministry of the Presidency.

Randall, A. 1981. Property entitlements and pricing policies for a maturing water economy. *Australian Journal of Agricultural Economics* **25**(3), 195–220.

RCEP (Royal Commission on Environmental Pollution) 1992. *Freshwater quality*. London: HMSO.

Rees, J. with S. Williams 1993. *Water for life: strategies for sustainable water resource management*. London: Council for the Protection of Rural England.

Ricardo, D. 1970. *On the principles of political economy and taxation*. Cambridge: Cambridge University Press.

Roberts, J. M. 1990. *The Penguin history of the world*. Harmondsworth: Penguin.

Robinson, J. & J. Eatwell 1974. *An introduction to modern economics*. Maidenhead: McGraw-Hill.

References

Ryle, G. 1949. *The concept of mind*. London: Hutchinson.

Sandmo, A. 1987. Public goods. In *The new Palgrave: a dictionary of economics*, J. Eatwell, M. Milgate, P. Newman (eds), 1061–1067. London: Macmillan.

Sawyer, M. 1994. Michal Kalecki. In Hodgson et al. (1994: 432–5).

Schiffler, M., H. Köppen, R. Lohmann, A. Schmidt, A. Wächter, C. Widmann 1994. *Water demand management in an arid country: the case of Jordan with special reference to industry*. Berlin: German Development Institute.

Schmid, A. 1994. Cost–benefit analysis. See Hodgson et al. (1994: vol. I, 104–108).

Shute, L. 1994. John Maurice Clark. See Hodgson et al. (1994: vol. I, 50–54).

Simmons, I. G. 1993. *Interpreting nature: cultural constructions of the environment*. London: Routledge.

Smith, A. 1961. *An inquiry into the nature and causes of the wealth of nations*. London: Methuen.

Söderbaum, P. 1994. Environmental policy. See Hodgson et al. (1994: vol. I, 193–9).

Spulber, N. & A. Sabbaghi 1994. *Economics of water resources: from regulation to privatization*. Dordrecht: Kluwer Academic.

van der Straaten, J. 1994. *The distribution of environmental costs and benefits: the case of acid rain*. Paper 94.09.037/2,Work and Organization Research Centre, University of Tilburg.

Stringer, D. 1995. Water markets and trading developments in Victoria. *Water* 22(1), 11–14.

Sunding, D., D. Zilberman, R. Howitt, A. Dinar, N. MacDougall 1994. *The costs of reallocating water from agriculture*. Berkeley: University of California Press.

Tellegen, E. 1981. The environmental movement in the Netherlands. In *Progress in resource management and environmental planning*, T. O'Riordan & K. Turner (eds), 1–32. Chichester: John Wiley.

Turner, K. & M. Postle 1994. *Valuing the water environment: an economic perspective*. Working Paper WM94–08, CSERGE, University of East Anglia.

UK Government 1994. *Sustainable development: the UK strategy* [ch. 8]. London: HMSO.

UNCED (United Nations Conference on Environment and Development) 1992. *The Earth summit*. Rio de Janeiro: United Nations.

Villarejo, D. & D. Runsten 1993. *California's agricultural dilemma: higher production and lower wages*. Davis: California Institute for Rural Studies.

Ward, R. C. & M. Robinson 1990. *Principles of hydrology*, 3rd edn. Maidenhead: McGraw-Hill.

WCED (World Commission on Environment and Development) 1987. *Our common future*. Oxford: Oxford University Press.

Webb, A. & C. Gossop 1993. Towards a sustainable energy policy. In *Planning for a sustainable environment*, A. Blowers (ed.), 52–68. London: Earthscan.

Willis, K. G. & G. D. Garrod 1990. *Valuing open access recreation on inland waterways*. ESRC Countryside Change Initiative Working Paper 12.

Winpenny, J. 1991. *Values for the environment: a guide to economic appraisal*. London: Overseas Development Institute.

— 1994. *Managing water as an economic resource*. London: Routledge.

Young, T. & P. G. Allen 1986. Methods for valuing countryside amenity: an overview. *Journal of Agricultural Economics* **37**, 349–64.

Zilberman, D., R. Howitt, D. Sunding 1993. *Economic impacts of water quality regulations in the San Francisco Bay and Delta.* Berkeley, California: Western Consortium for Public Health.

INDEX